The Syrian Revolution

The Syrian Revolution

Between the Politics of Life and the Geopolitics of Death

Yasser Munif

First published 2020 by Pluto Press
345 Archway Road, London N6 5AA

www.plutobooks.com

Copyright © Yasser Munif 2020

The right of Yasser Munif to be identified as the author of this work has been asserted by him in accordance with the Copyright, Designs and Patents Act 1988.

British Library Cataloguing in Publication Data
A catalogue record for this book is available from the British Library

ISBN 978 0 7453 4071 5   Hardback
ISBN 978 0 7453 4072 2   Paperback
ISBN 978 1 7868 0580 5   PDF eBook
ISBN 978 1 7868 0582 9   Kindle eBook
ISBN 978 1 7868 0581 2   EPUB eBook

Please note, some maps are drawn from Wikipedia and amended by a Syrian cartographer who wishes to remain anonymous for political reasons.

Every effort has been made to trace copyright holders and to obtain their permission for the use of copyright material in this book. The publisher apologises for any errors or omissions in this respect and would be grateful if notified of any corrections that should be incorporated in future reprints or editions.

Typeset by Stanford DTP Services, Northampton, England

To Syrians who rose up against tyranny and injustice.

Contents

Acknowledgments — viii

Introduction — 1
1. Necropolitics: The Taxonomies of Death in Syria — 10
2. The Geography of Death in Aleppo — 43
3. Nation Against State: Popular Nationalism and the Syrian Uprising — 95
4. The Politics of Bread and Micropolitical Resistance — 120
5. Participatory Democracy and Micropolitics in Manbij: An Unthinkable Revolution — 144
Conclusion — 157

Notes — 159
Index — 189

Acknowledgments

Writing this book was a collective endeavor and I wouldn't have been able to achieve it without the help of many friends and colleagues. They are too numerous to mention here but I am deeply grateful to each one of them for their invaluable and brilliant suggestions to previous drafts of chapters included in this book.

An earlier version of the fifth chapter, "Participatory Democracy and Micropolitics in Manbij: An Unthinkable Revolution" was previously published in the book *Arab Politics beyond the Uprisings: Experiments in an Era of Resurgent Authoritarianism*, ed. Thanassis Cambanis and Michael Wahid Hanna (New York: The Century Foundation Press, 2017). I would like to thank the editors for granting the permission to print it here. I am also grateful to the Arab Council for the Social Sciences for funding part of the research for this project.

This book would not have been possible without the generous help of friends in Manbij, Raqqa, Aleppo, Damascus and elsewhere in Syria who challenged my ideas about Syria and taught me to unthink everything I knew about my country to begin to understand revolutionaries' praxes and ideas. My deep gratitude and appreciation goes to them.

I began working on this project almost five years ago and it went through many iterations. I am deeply grateful to my extraordinary wife, Elsa Wiehe, who shared every step of this journey. Her constant support, critical questions, and encouragements have made writing this book possible. I would also like to thank my wonderful daughters, Yara and Zoya, for their patience and despite their multiple acts of sabotage. I hope that the day they open this book to read it, dictatorship will have become something of the past.

Introduction

In March 2015, the World Social Forum (WSF) was held for the second consecutive time in Tunis. The Tunisian capital was chosen by the organizers to emphasize the importance of the Arab revolts and their impact on social movements globally. Evidently, there were many panels and events about the Arab and Syrian uprisings. During the Forum, Assad supporters attacked several dozens of people attending a panel about the Syrian revolt that I helped organize. As I finished introducing one of the speakers, the late Syrian-Palestinian intellectual Salameh Kaileh, about 20 men entered into the conference room causing a scuffle. They chanted pro-Assad slogans, and removed pro-revolution posters and a banner from the walls. When attendees tried to steer them out of the room, they became threatening and shouted, "You are terrorists! No to regime change in Syria!"

Two years prior, a similar incident had taken place at the same Forum. Assad supporters became quickly agitated when they realized that half a dozen of us were collecting signatures for a petition that condemned the Syrian regime's atrocities.[1] A man in his 30s was louder than the others and was pushing for a confrontation. "There is no revolution in Syria," he shouted, "there is only a global conspiracy against the only anti-imperialist state in the Arab World and we will defeat it!" He was probably a provocateur but many of the people with him were persuaded that all protests against Assad were the result of Western manipulations. Some argued that Gulf countries paid infiltrators to destabilize the regime and justify an attack on, and later an invasion of, Syria. Within minutes, approximately 200 pro-Assad supporters and pro-revolution activists were facing each other, and ready to clash. The WSF security and organizers spent close to an hour trying to defuse the tension. Later that day, several people physically assaulted a friend who was on her way to attend a different event.[2] Heated arguments about the Syrian revolt and confrontations such as the ones described here are not unusual; in fact, they have become the norm since 2011. In the United States and Europe, it is common to see pro-Assad supporters or pro-revolution activists picket or disrupt the events of their political rivals.

These conflicts have also become an integral part of the academic landscape. Sheffield University has recently asked Professor Piers Robinson to leave his post as a result of a campaign that exposed his affinity to autocratic regimes such as Russia, and use of conspiracy theories regarding Syria.[3] The intense confrontation in academia is not confined to the social sciences, political science, or history departments. There are fierce debates in areas that are usually impervious to politics. For example, a few years ago, environmental scientists took part in a discussion about the role of intense, multi-year droughts that preceded the 2011 revolt. Some scholars argued the droughts caused the displacement of the population on a massive scale leading to the formation of poverty belts around large Syrian cities such as Aleppo and Damascus. In their view, these structural processes were instrumental in generating protests against Assad.[4] Others contended that decades of mismanagement of water resources, combined with unrestrained irrigation, played a more important role in the pauperization of the peasantry.[5] Scholars and professionals working in health had a heated debate about their ethical role in the Syrian conflict. Dr. Annie Sparrow wrote an article about the World Health Organization (WHO)'s criminal complicity with the Syrian regime. While most international health institutions, organizations, and practitioners took a firm stance against the Syrian regime, others such as the WHO allowed it to weaponize the medical field against its population.[6]

Another fervent debate took place in the field of chemistry and ballistic science to determine who was behind the various chemical and non-chemical attacks in Syria. Theodore Postol, professor of science, technology, and international security at the Massachusetts Institute of Technology is one of the leading experts in the field.[7] He has become the academic of reference when regime loyalists and allies seek an authority figure in the field of ballistic science to absolve Assad of inconvenient chemical attacks.[8] More serious scholars, however, have debunked his claims by highlighting the many inconsistencies in his analyses.[9]

In the same vein, the archeological community is split around the role of the Assad regime in the destruction and looting of archeological sites such as Aleppo's Old City, Palmyra, or Krak des Chevaliers castle.[10] Some professionals in the field played a significant role in undermining the Syrian revolt using their credentials and expertise. For instance, when Russian and Syrian forces defeated ISIS in the Syrian Desert, Vladimir Putin contacted the then Director-General of UNESCO, Irina Bokova,[11] to inform her about "the successful completion of military operations to liberate the city of Palmyra, one of the treasures of world cultural heritage."[12] It is logical that the Russian president put the blame on ISIS for the destruction of Syria's historical sites but that UNESCO is helping him to polish his image is problematic to say the least.

Maamoun Abdulkarim, the Director-General of Antiquities and Museums in Syria, argued that the greatest threat to Syria's archeology is ISIS. He told Ian Black, the *Guardian*'s journalist, that the Syrian government has been protecting archeological objects during the war. He then concluded, "[n]ow we hide things in Damascus, but if Damascus falls, what can we do? We are not supermen."[13] The *Guardian*'s article describes the atrocities of ISIS in the ancient city of Palmyra but fails to push back against the government's narrative according to which it is safeguarding archeological sites and artifacts. It mentions the tragic killing of Syrian archeologist Khaled al-Asaad by ISIS but doesn't make any reference to Tadmur prison, which is within walking distance from the archeological site and which witnessed the massacre of more than a thousand prisoners on June 27, 1980. The journalist briefly explains, "irreparable damage has been done by government forces as well as their enemies—in the battle for the Crusader castle of Krak de Chevaliers and the magnificent 14th-century Madina souq in old Aleppo."[14] However, the thrust of the article is about ISIS's systematic destruction of archeological sites and the government strategies to prevent that. Black writes,

> [i]n scenes that echoed the second world war story of the Monuments Men who saved artworks from destruction or looting, 24,000 objects were also brought by truck from Aleppo earlier this year. The heroine of that operation was a 25-year-old Syrian archaeologist who stayed with the army-escorted convoy for a hazardous 11-hour journey.[15]

The irony is that *The Guardian* published an article, *less than a year later*, titled "Syrian Troops Looting Ancient City Palmyra, says Archaeologist."[16] Finally, Black neglects to explain that the regime reduced entire neighborhoods into rubble and ruins, and destroyed large sections of cities where countless civilians perished and archeological sites were annihilated. This article is paradigmatic of mainstream coverage of Palmyra and other archeological sites in Syria.[17] The main battle for these journalists is between Western civilization and the Islamic State's barbarism; their coverage of Palmyra, unfortunately, is the by-product of a Eurocentric world-view.[18] This explains why many believe that the Syrian regime is a legitimate partner when it comes to defeating ISIS and preserving world historical sites. Maamoun Abdulkarim understands these dynamics and knows how to exploit Western sensibilities as his conversation with Black demonstrates.

The tragic reality is that the debates occurring within the field of archeology are not exceptional or surprising. Scholars in fields as diverse as health, environmental science, ballistic science, not to mention the social sciences, history, or the humanities are all deeply split around the same issues. Many scholars decided to become guardians of a hegemonic geopolitical order instead of mobilizing resources within their fields to support Syrians' struggle for justice and self-determination. Conversely, these professional circles should use the Syrian revolt to decolonize their disciplines, and make sure they're not simply vehicles of state power or international law deployed against oppressed groups in Syria. These seemingly impartial academic or technocratic debates cannot be delinked from the lives of Syrians and their daily experiences. As seen above, some of these debates have lethal consequences, while certain academics and professionals are complicit in the regime's crime.

In this book, I argue that the deep divide described here transcends any specific field and should be understood as a confrontation between two world-views. The first one focuses mostly on the macro-political dimension and utilizes conventional theoretical tools, such as the "nation-state," "international relations," or "geopolitics," to analyze the Syrian crisis. The second is attentive to micropolitics and everyday practices; it avoids simplistic labels that might invisibilize the people struggling on the ground. Simply put, the first reinforces the "politics of death," while the second produces a multifaceted "politics of life." The book proposes an unconventional reading of the Syrian revolt by analyzing the anatomy of violence and grassroots struggles. It argues that the conflict between revolutionary forces and the Syrian regime should be understood primarily as a confrontation between micropolitics of life and macropolitics of death. It is a clash between authoritarian technologies of power and grassroots micropolitics.

THE QUESTION OF METHODOLOGY

The Arab revolts are a pivotal moment in the history of the region and require innovative conceptual frameworks for their analysis. While the scholarships of Edward Said and Frantz Fanon have altered the politics of knowledge production about the Arab and Muslim worlds, Orientalist tropes about the region remain well entrenched. Below, I identify three dominant theoretical frameworks that are frequently deployed to examine the Syrian conflict. Each, in its specific way, is reproducing the old structures of domination and reinforcing Western epistemologies. This is not a proposition to simply reject them but rather to rethink them. The first one encompasses various Orientalist approaches that essentialize identities, cultures, and histories. It proposes totalizing ideas about religion, politics, and gender relations. Orientalist narratives in the twenty-first century have become subtler compared to what they were a few decades ago. Many conservative think-tanks in the West still operate within the confines of these archaic frameworks about the Middle East and the Arab world. They reassert essentialist assumptions about tribes, sects, and gender relations. They privilege the military aspect of the conflict with a focus on terrorism, fundamentalism, and sectarianism.[19]

The second approach maintains the nation-state at the center of the analysis. When the state is not problematized, its violence is invisibilized and legitimated. Geopolitics is one such theoretical framework that prioritizes the state at the expense of other actors. From that standpoint, the state is the unit of analysis and anything that operates beneath or above it is dismissed. The main issue is that this perspective essentializes the state and presents it as an ahistorical entity. Geopolitics emphasizes the state and its interests but erases popular struggles because they do not use the same language nor do they operate on the same plane. Geopolitical analyses, while important, cannot fundamentally challenge the world-system or international relations. More recently, feminist geopolitics,[20] or anti-geopolitics,[21] shifted the focus to micropolitics and to the agency of marginal actors but this new scholarship is marginal in the fields of geopolitics and international relations.

A third strand of scholarship revolves around New Social Movement (NSM) theory. While NSM could be insightful in a Western context, it does little in a non-Western one such as the Arab World. In Syria, where anti-politics is the norm, social movements do not have a space to operate in. The idea of building a social movement to counter certain policies was simply inconceivable in Assad's Syria, where dissident political actions were harshly repressed. Thus, the prison system became a central institution in Syria. The work of Iranian-American scholar Asef Bayat, especially his book *Life As Politics*, is an important contribution in that regard. He develops non-Eurocentric concepts that take into consideration the history and specificities of Arab and Muslim societies.[22]

The book develops a heuristic analysis that operates against these frameworks. It contends that such an approach is necessary in the context of the Arab revolts where Western dominant perspectives are ineffective for the reasons explained above. The different chapters engage with intellectuals as diverse as Giorgio Agamben, Achille Mbembe, Gilles Deleuze, and Frantz Fanon to explore new paradigms. Finally, the following chapters avoid traditional linear historical narratives and instead propose a Nietzschean genealogy. The book investigates the ways bodies and institutions are enmeshed in webs of power and are the result of political, cultural, and economic conflicts. It rejects a history that consists of uncritically cataloging events from the past. Friedrich Nietzsche explains in "On the Uses and Disadvantages of History for Life" that there is need for several types of history. Monumental history is essential because it reminds us of human capabilities and provides examples of compelling historical figures. Antiquarian history provides a sense of communities and deep roots in a shared past. Finally, critical history requires the forgetting of specific moments, the repression of others, and also a selective appropriation of the past.[23] Following Nietzsche, French philosopher Gilles Deleuze notes that the task of the historian is to describe a *becoming* rather than fixed events in the past. "Becoming" is the result of conflicting forces that are in flux. In that context, the Syrian revolt opens a new space that allows for revisiting and altering the history of the present. The chapters below engage with various theoretical frameworks but they all attempt a genealogical reading of the past. History is therefore constantly reconstructed from the standpoint of the present rather than being a collection of static facts from the past. In Chapter 3, for example, I argue that popular nationalism that the Syrian revolt produced in 2011 could be read as re-enactment of nationalism of the early 1920s as well as a rejection of official nationalism of the 1960s onward. Likewise, Chapter 1 shows that participatory democracy, which re-emerged in 2011, cannot be delinked from Syria's democratic years of the 1950s. Overall, I contend that there are moments where the politics of life operates through historical resonance rather than temporal continuity. The purpose of a non-linear reading is primarily to deconstruct the official narrative, and revisit moments in Syria's history that pre-date Assad's era.

I began the research for this book in 2011 when I visited Syria for several weeks. I spent two months in 2011 and 2012 in regime-controlled territories. In 2013 and early 2014, I spent close to three months in Northern Syria in areas controlled by the Free Syrian Army and civilian groups. I was mostly based in Manbij, which is a medium-sized city located in northeastern Aleppo. I also visited Raqqa, Aleppo, and other smaller cities, as well as several camps in Syria and Turkey. In these different locations, I conducted close to 200 interviews with activists, notables, youth groups, doctors, fighters, and internally displaced persons. After ISIS's takeover of Manbij and other cities in Northern Syria in 2014, it became impossible to return to these regions. Between 2014 and 2016, I spent time in Gaziantep and other Turkish cities close to the Syrian borders. During these trips, I conducted interviews with activists and refugees who had fled the violence of the regime and ISIS. While the ethnographic component in this book is marginal, my stay in Syria and interviews with Syrians from various backgrounds informs each one of the chapters.

THE POLITICS OF LIFE AND DEATH

The first two chapters investigate the devastating implications of macropolitics. They analyze the geography of violence deployed by the regime to crush revolutionary forces. They unpack the concept of macropolitics in the context of the Syrian revolt. They show that macropolitics is often undemocratic, destructive, and counter-revolutionary. It is no coincidence that regional powers, Western elites, as well as international institutions would choose a macropolitical lens to apprehend the Syrian conflict. By doing so, these actors deliberately chose to sideline revolutionary struggles and consequently empower the Syrian regime. These chapters examine the multidimensional strategies utilized by the regime and other hegemonic actors to undermine grassroots resistance.

The following chapters propose a micropolitical analysis of the Syrian uprising by focusing on Manbij. Grassroots resistance is by definition local, heterogeneous, and complex. This suggests that only an exploration from below could examine the multifaceted revolutionary praxes. It starts with an analysis of popular nationalism since 2011. The main purpose of emergent nationalism is to counter the despotic ideology of the Baath party. In addition, these chapters analyze micropolitical processes in the liberated territories and the challenges revolutionaries face on a daily basis. Micropolitics can take many forms including the management of the city, the distribution of bread in war-torn territories, and underground resistance in regions controlled by regime forces. Overall, the book proposes two alternative readings of the events in Syria since 2011. The first one revolves around state violence and its destructive impact, while the second suggests that a politics of life could shed light on micro-processes that are mostly invisible to external observers. Finally, it shows that the politics of life can emerge from within the cracks of the geopolitics of death.

Using the concept of "necropolitics," Chapter 1 proposes a taxonomy of death in Syria since 2011. Death in its various manifestations is a crucial entry point to understanding the anatomy of violence. The Syrian regime employed various technologies of death including starvation, torture, siege, indiscriminate bombing, chemical attacks, massacres, assassinations, etc… For a distant observer, the lethal violence appears indiscriminate but a thorough analysis shows there is a rationale beneath the orgy of death in Syria. The analysis demonstrates that each type of death (slow/quick; sensational/ordinary; random/calculated…) serves a specific purpose. In the end, the regime's success in crushing the revolt, at least temporarily, resides in its ability to continually optimize and carefully calibrate its lethal techniques. Chapter 2, "The Geography of Death in Aleppo," provides an in-depth study of urban destruction (urbicide) in Aleppo, the second largest city in Syria. Revolutionary forces liberated eastern Aleppo in July 2012 and controlled it for more than four years until its fall in December 2016. The city offers a strategic site to examine the uneven geography of violence as well as its scale and intensity. The analysis focuses on the Syrian regime's utilization of spatial strategies to subdue the population and weaken grassroots struggles. The regime's urban warfare is informed by Aleppo's history as well as its social and demographic compositions. For example, the location of fixed checkpoints as opposed to mobile ones was carefully determined. The aerial maps show that despite their inaccuracy, barrel bombs were dropped on specific areas while chemical and clusters bombs targeted others. Urbicide in Aleppo represents a microcosm of the war that is still raging in the rest of Syria.

Chapter 3, "Nation against State: Popular Nationalism and the Syrian Uprising," examines nationalism in the context of the Syrian revolt. Through their struggles for freedom and self-determination, Syrians produced a new form of popular nationalism. This nationalism was dominant at least during the first two years before Islamic forces became hegemonic. It was deployed during the uprising as a tool for cultural resistance against the regime's instrumentalist and despotic nationalism. Simply put, official nationalism is scripted, monolithic, and rigid. Popular nationalism is elusive, performative, and plurivocal. To justify its grip on power, the Baath party weaponized nationalism for more than half a century. The cultural clash between these two nationalisms is the logical result of the military confrontation that is happening on the ground. Chapter 4, "The Politics of Bread and Micropolitical Resistance," examines the geography of bread in Syria since 2011 using two approaches. The first approach explores the political economy of bread with a focus on state policies and economic programs under the Baath rule. It analyzes the ways the state handled wheat shortages and prevented bread riots in the past. The Syrian regime used bread as a lethal weapon during the revolt to suppress opposition and undermine grassroots resistance. In the second part of the chapter, the focus shifts to the minutiae of everyday resistance and the ways the residents in Manbij developed new strategies to produce, transport, and distribute wheat. These strategies were essential for the survival of the population and the continuation of the revolt. An analysis of the geography of bread in war-torn Syria allows for a better understanding of grassroots politics. Finally, Chapter 5, studies micropolitical struggles and local governance in the liberated territories using Manbij, a city in Northern Syria, as a compelling example of successful grassroots governance during the two-year period between the Syrian regime's withdrawal from the city in 2012 and the Islamic State's takeover in 2014. During this interregnum, revolutionary forces reconfigured the city from the ground up by creating inclusive spaces, forming horizontal networks, and building democratic institutions. A micropolitical analysis of the revolt reveals a complex reality where military confrontation plays a peripheral role. Everyday resilience and unrelenting organizing constitute the backbone of the Syrian revolution. Revolutionary forces in Manbij overcame important challenges such as the lack of resources or the violence of the regime and built a vernacular form of democracy. Through experimentation, they created new institutions to transform their city and make it livable. The study shows that reality in the liberated regions is much more complex than the way it is represented in the media or in public discourses.

1
Necropolitics: The Taxonomies of Death in Syria

The silence of slippers is more dangerous than the sound of boots
French priest Martin Niemöller

It's not a civil war. It's a genocide. Leave us die but do not lie.
Kafranbel banner, November 2, 2012

This chapter explores the taxonomies of death and technologies of violence that the Syrian regime has deployed over the past eight years to crush the uprising. It argues that the current politics of death would not have been possible without the imposition of a state of emergency in Syria. Emergency was a vital political tool that allowed the regime to maintain power for several decades. It would not have been consequential, however, without the prison system and state terror that enforce it. The chapter begins with a brief history of the state of emergency and how Assad used it to eliminate his political opponents and consolidate state power. It explores the significance of Giorgio Agamben's "state of exception" in the Syrian context. The second section examines the various ways the politics of death or necropolitics was implemented during the Syrian revolt. Using Achille Mbembe's notion of necropolitics as an entry point, I argue, can shed light on new aspects of the conflict. Finally, in the last section, I explore the prison system in Syria and the ways it allowed the regime to conceal its atrocities.

STATE OF EMERGENCY SINCE 1963

The Baath party takeover in 1963, followed by Hafez al-Assad's authoritarian rule after 1970, aborted the democratic process that Syrians had built in the 1950s. Political life in the mid-1950s was organized around two poles: on the one hand, the hegemonic politics of urban notables, and on the other hand, the pan-Arab and progressive parties. During this democratic period, Syrians had first-hand experience with political pluralism, free elections, and freedom of the press.[1] The landed oligarchy, which had dominated the political, economic, and cultural spheres until then, was losing power. Instead, Arab nationalists such as the Baath party and Nasserists, as well as progressive parties, were gaining momentum.

The Baath party first emerged as a political force in the Syrian parliament, thanks to the democratic environment of the mid-1950s. The parliamentarian progressive coalition, to which the Baath belonged, was unable to implement land redistribution and other reforms due to the landed oligarchy's powerful opposition. The parliamentary members of the urban notables won the elections and undermined the ambitious program. As a result, the Baath party mobilized workers and students in the street, and attempted to control crucial positions in the army.[2]

Baathists were also facing another important challenge in the 1960s, namely, the Soviet Union's growing influence in the Middle East. In 1957, the Baath party ended its alliance with the Syrian Communist Party, fearing that their growing popularity and ability to mobilize the working class would sideline the pan-Arab party. By the end of the year, the landed oligarchy and the Baath both felt threatened by communists, who had the support of the Soviet Union. In the context of the Cold War, Syria became a site of conflict between the United States and the Soviet Union, and the escalation between the two was on the verge of taking a military turn. The Baath party believed the best way to regain influence in the political sphere was to forge a union with Abdel Nasser, who shared their ideal of pan-Arabism.

The United Arab Republic (UAR) was established in February 1958, and was primarily a bureaucratic entity that maintained Nasser's dominance over Syria's political life. Nasser banned Syrian parties and controlled every aspect of politics. He ruled Syria with the help of traditional politicians, while purging the most progressive officers in the army. His nationalization of banks and the implementation of the agrarian reform alienated the same political class he was allied with. During this tense political conjuncture, a conservative segment of the army seized power through a military coup in September 1961 and put an end to the UAR.[3]

The failure of the union with Egypt and the military takeover by the Separatist Officers split the Baath party into two main competing branches. The first was led by the ideological founder of the party, Michel Aflaq, but was marginal and mostly based in Damascus. The second was dominant in cities such as Latakia, Deir ez-Zor, and Dara'a, and was backed by a clandestine military committee constituted of Baathists and Nasserists. While the military branch of the Baath was critical of the union with Abdel Nasser, since he had banned their party during the union, they nonetheless formed an alliance with Syrian Nasserists for opportunistic reasons. During the 1960s, Nasserists were influential in the entire Arab world, and especially so in Syria. The Baath would not have been able to overthrow the Separatist Officers and seize power in 1963 without the vital support of the Nasserists. Once in power, the coalition marginalized the conservative notables who backed the Separatists.[4] The Baathist and Nasserist alliance was initially motivated by radical ideas, and opposed the powerful notables who had dominated political life until that point. Baathists operated in an anti-imperialist context, where opposition to Western interests in the region and solidarity with Palestinian struggle were the norm.

Akram al-Hourani, an Arab populist from Hama who had played a prominent role in opposing the hegemony of urban notables, merged his party with the Baath in 1952. He advocated for an ambitious program of land reform, and brought many peasants to the Baath who later became the pillars of the party. He went into exile when the Baath seized power in 1963. After its successful coup in March, the Baath consolidated power through the military and security apparatuses, but also by mobilizing its peasants and workers bases.

The Iraqi Baath party, which had led a successful coup a few weeks earlier, incentivized the Syrian Baath to go ahead with its own coup. A military committee that included three Baathists (namely, Muhammad Umran, Salah Jadid, and Hafez al-Assad) in addition to two Nasserists planned and executed the coup. Michel Aflaq, the Baath secretary-general, reluctantly approved the plans, despite his disapproval of the military's increasing influence in party politics.

After the coup, Aflaq and his followers were marginalized, while the progressive leaders of the party who favored socialist policies dominated the political scene. They began implementing their program of land reform and industrial development, while at the same time marginalizing the urban notables who owned most of the land. Nasser began land redistribution in 1958, but the radical branch of the Baath party, which passed land reform laws in 1963 and 1966, mostly implemented it. Land reform was completed in 1970 when Assad came to power.[5] Between 1963 and 1970, internal conflicts and repeated political purges within the Baath led to the gradual domination of the pragmatic branch that Assad represented. Salah Jadid and the radical tendency he headed were subsequently frozen out, especially after the 1967 War and Syria's defeat by Israel. Hanna Batatu, a renowned Palestinian historian, explains, "[w]hen the Six Day War of 1967 broke out, [Assad] was still a military amateur and did not have the qualifications to be the thinking head of the armed forces."[6] According to him, on June 10, 1967, Assad issued the devastating Communiqué No. 66, in which he announced the fall of Qunaytara when it was actually still in the hands of the Syrian army. The announcement caused panic at the front line, and did in fact precipitate the fall of the city. To protect the regime, he pulled out the 70th armored brigade to reposition it nearby Damascus. After the humiliating defeat against Israel and the loss of the Golan Heights, Assad blamed Jadid, despite being Minister of Defense himself and playing a central role in the war. In 1970, he finally threw Jadid in jail, consolidated his rapprochement with segments of the Syrian bourgeoisie, and put a halt on the socialist program of the Baath party.[7]

The state of emergency

To maintain power, Hafez al-Assad built a coup-proof regime that survived several serious crises after 1963. The internal conflicts of the Baath party, in addition to the turbulent political life of the 1950s and 1960s, informed his trajectory. He implemented a complete restructuring of the Syrian state to prevent military coups. Assad built a patrimonial state where the upper echelons of officers were recruited from the Alawite sect. The vast majority of these officers were loyal to the regime, despite a few defections in times of crisis. However, as political scientist Steven Heydemann explains,

> A second resource played a critical role in stemming opposition advances and stabilizing the regime: informal networks of nonstate actors, organized on the basis of familial ties, sectarian affinity, or simple mercenary arrangements, and cultivated by regime elites over the years to provide a range of (often illegal) functions that could be conducted without any formal scrutiny or accountability.[8]

The regime's ability to survive the popular opposition is built on solid foundations. According to Syrian scholar Radwan Ziadeh, Assad's authoritarian rule was based on a triangle of power, namely, the security branches, the army, and the Baath party.[9] It was these dense networks of power that allowed the regime to crush the rebellion. However, Ziadeh's framework ignores class analysis, which is essential to understand Assad's consolidation of power. The modern history of Syria shows that Assad paid close attention to class composition inside Syrian society, and built a complex balance between elites and popular classes.[10]

One of the most important technologies of oppression in Assad's regime was the state of emergency. The Separatist Officers first imposed it in 1962, and the Baath party maintained it until 2011. It allowed the regime to operate at the periphery, and often outside the legal system, without any significant backlash. Under the state of emergency, the Syrian regime incarcerated thousands of its political opponents and assassinated many in Syria and abroad without accountability. This was the case for Salah el-Din al-Bitar, Muhammad Umran, Kamal Jumblat, and Rafic al-Hariri among others.[11]

The codification of death

The Separatist Officers, who seized power after the failure of the union with Egypt, proclaimed a state of emergency on December 22, 1962. When the Baath party took power in the following year, the first decree it issued on March 9, 1963 was concerning the amendment of the state of emergency and the imposition of martial law. The state of emergency had been issued in Syria multiple times before 1962. Edward Ziter explains, "[t]hese laws replaced the Emergency Laws instituted by Nasser in 1958 during Syria's short-lived union with Egypt. Before that, martial law had been instituted in both 1953 and 1956."[12] The state of emergency was declared several more times before the 1950s. As the French prepared to occupy Syria, Yousef al-Azmeh imposed martial law in July 1919, a few days before the unsuccessful battle of Maysaloun.[13] The French colonial power used martial law multiple times, starting in 1920, when it faced a popular rebellion. It was implemented in the rebellious regions only, namely, Damascus, Hawran, and Jabal al-Druze.[14] The state of emergency was declared again in 1939 at the outbreak of World War II, as the French government prepared for the war.[15] In 1948, the Syrian government declared it, after riots erupted in Damascus and other cities due to a heated debate in parliament about the partition of Palestine in the previous year.[16]

Before 1962, the state of emergency was typically lifted after a few months or years. When Assad seized power, he maintained it for almost half a century. It was finally lifted in April 2011, a few weeks after the eruption of the revolt. The end of the state of emergency and the release of political prisoners were the protesters' main demands during the first months of the revolt. The new legislation put an end to the state of emergency and also dissolved state security courts.[17] The state of emergency was hastily replaced by a counter-terrorism law, which went into effect on July 2, 2012.[18]

For five decades, any criticism of the Syrian regime was severely punished under the pretext that it could lead to the "weakening of national sentiment."[19] In March 1968, the Syrian government created several courts and new laws to consolidate its power: the Supreme State Security Court (SSSC), martial courts, and military field courts. They were responsible for enforcing emergency laws and operated outside the legal and court systems; as such, they had far-reaching power and only responded to executive orders. There was no mechanism to appeal their decisions. In addition, several articles were added to prevent the prosecution of officers involved in the disappearance, torture, or killing of Syrians.[20]

The new constitution went into effect after being approved by the people's assembly in 1973, and was valid until February 2012, when the Assad regime, under popular pressure, amended it.[21] The 1973 constitution gave unprecedented powers to the president. Article 101 states that: "The President of the Republic can declare and terminate a state of emergency in the manner stated in the law."[22] Article 107 explains, "[t]he President of the Republic can dissolve the People's Assembly through a decision giving the reasons."[23] Moreover, Article 111 gives power to the president to assume authority when the People's Assembly is not in session and "even when the Assembly is in session if it is extremely necessary."[24]

The international context allowed the Assad family to impose the state of emergency for almost half century with no real consequences. International institutions such as the United Nations failed to address the issue, due to the pressure of powerful state actors. Many states were reluctant to voice opposition to the state of emergency, in the event they needed to implement it themselves. In addition, in the 1970s and 1980s, many states around the world declared states of emergency and utilized them to repress their own societies. In 1978, 30 countries had imposed a state of emergency; whereas by 1986, their number had grown to 70.[25] Thus, the international politics of silence allowed the Syrian regime to maintain the state of emergency without being held accountable. In 1979, Amnesty International published its first report about human rights violations in Syria, addressing the question of torture and the state of emergency.[26] The report came out 17 years after the emergency was first implemented and didn't have much impact.

Throughout the Assad rule, the state of emergency was an effective tool for repressing political opponents. The Seventh Country Conference of the Baath Party in 1979 concluded that there was a need for "intensifying the political security campaign to eradicate the gang of the Muslim Brotherhood and its foundation in the state and society."[27] To justify and consolidate the state of emergency, the regime criminalized the Muslim Brotherhood and presented the group as an eminent and permanent threat. In 1980, the Assad regime issued the infamous Law 49, which banned the Muslim Brotherhood and made affiliation to their party punishable by death. After crushing the Brotherhood's rebellion in 1982 and the killing and incarcerating of thousands of political opponents, the regime was able to impose a ubiquitous politicide for several decades. It was only after the death of Hafez al-Assad in 2000 that political opposition re-emerged.

The state of emergency is often discussed in the context of international law and human rights treaties.[28] The studies that analyze it in the Syrian context are of two types. The first is based on international law,[29] while the second examines the Syrian context through a human rights prism.[30] Both fields are legalistic in nature, examining the tension and contradictions between the Syrian constitution and international law. They emphasize the contradictions within the Syrian legal system, and denounce the state's violations of human rights under the state of emergency. Legal scholar Michael Macaulay, for example, explains that the state of emergency in Syria did not meet the principle of legality in accordance with domestic or international laws, and as such was in clear violation of both, and therefore should have been abolished on that basis.[31] While the legal framework is useful, it has important limitations. It does not question the legitimacy of the state or the sovereign, or explain the original violence embedded in law. To address these issues, I explore the implications of the state of emergency beyond the legalistic framework by invoking the work of Italian philosopher Giorgio Agamben.

THE SOVEREIGN AND STATE OF EXCEPTION

Agamben's innovative take on the state of emergency and the sovereign could shed light on the mechanisms of despotism in Syria. The Italian philosopher examines the implications of the state of emergency when it becomes the rule rather than the exception. As such, his work is particularly relevant in Syria, a country where the "state of exception" was imposed for almost half a century. In the first pages of his book *State of Exception* (2005), Agamben notes:

> The state of exception tends increasingly to appear as the dominant paradigm of government in contemporary politics. This transformation of a provisional and exceptional measure into a technique of government threatens radically to alter—in fact, has already palpably altered—the structure and meaning of the traditional distinction between constitutional forms.[32]

To examine the significance of the state of emergency, Agamben engages with the work of two German intellectuals. He builds on theories of cultural critic and radical thinker Walter Benjamin, while criticizing the stern conceptual framework of conservative political theorist Carl Schmitt.

Schmitt has written extensively about the sovereign and the state of emergency. He argues that the absolutist sovereign is he who makes the exception. For him, the main attribute of the sovereign is his power to decide and impose a state of exception. The sovereign should therefore be understood as the power to suspend the legal when needed, and restore it when necessary. Schmitt explains that the state should be able to suspend law to protect its sovereignty. The state, which is usually the enforcer of law, is in some cases the *force* of law. For him, it can declare the state of emergency simply because it is sovereign.[33] In *The Guardian of the Constitution* (2015), Schmitt argues that the head of the executive power, not the legal branch, should be the guardian of the constitution.[34] According to Schmitt, the difference between constituted law and the sovereign who enforces the law is irrelevant when the state of exception is declared. The only thing that governs the sovereign and puts limits on it is the sovereign itself. Whenever the old order is overthrown and the new one is established, the law is suspended. During this transitional period, Schmitt argues, only the sovereign can decide what the content of law is.

What makes the sovereign unique is his/her ability to be at once inside and outside the legal order. Critical theorist, Yehouda Shenhav explains, "Instead of the legal rule of law, Schmitt suggested that 'Sovereign is he who decides on the exception', and that the exception in jurisprudence is analogous to the miracle in theology."[35] That is precisely what allows the sovereign to order the state of emergency and to rescue the state from collapsing. Schmitt was particularly dissatisfied with European liberal democracies that could not develop an adequate response when facing political crises in the 1920s. He argues that the state of exception and the exceptional power of the sovereign are vital to rescue the state from complete collapse. As such, Schmitt believes the only reason the sovereign transgresses the rule of law is the public good.[36]

Agamben provides a radical reinterpretation of Schmitt's concept of sovereignty. He explains that when the sovereign suspends the rule, he operates in an in-between space that is both legal and non-legal. The state of exception blurs the line between the legal and non-legal, between private and public, between the political and the juridical. Agamben sees the state of emergency as indeterminate, especially since World War I. Unlike Schmitt, who believes the state of emergency is a temporary event, Agamben shows that in the twentieth century, it has become the status quo. Following Benjamin, Agamben explains that the state of emergency is not the exception but rather the rule in the contemporary world. In addition, Agamben examines it not only from the standpoint of the sovereign but, more importantly, from the perspective of the subject who experiences it. The state of exception needs to produce an outside or an enemy. As a result, the law applies to those within, that is, those who belong to the polity. The ones left out become outlaw and are denied any legal rights; they are reduced to bare life.

Bare life is the condition of living at the edge of the juridico-political space. They are often undocumented, migrant, racialized, sexualized, poor, or "abnormal" individuals. Agamben warns,

> [As] long as the two elements (law and anomie) remain correlated yet conceptually, temporally and subjectively distinct, their dialectic can nevertheless function in some way. But when they tend to coincide in the single person, when the state of exception, in which they are bound and blurred together, becomes the rule, then the juridico-political system transforms itself into a killing machine. The normative aspect of law can thus be obliterated and contradicted with impunity by a gov-

ernment violence that—while ignoring international law externally and producing a permanent state of exception internally—nevertheless still claims to be applying the law.[37]

To overcome the violence of law and counter the power of the sovereign, Agamben introduces Benjamin's notion of revolutionary violence. The German cultural critic makes a distinction between two forms of violence. The first is the "[l]aw-preserving violence [that] pertains to the order of 'constituted power' and consists in legal, para- or extra-legal measures that sustain the existing law and order of things." The second Benjaminian violence is revolutionary and operates outside the law, instituting the 'constituent power.' Only the second type of violence, according to Benjamin, has legitimacy and cannot be controlled by the sovereign.[38]

Benjamin's second type ruptures the link between life and the legal. Through revolutionary violence, the life of the revolutionary subject is no longer bound by the legal system. For Benjamin, the only violence that can oppose the violence of the sovereign is pure or revolutionary violence. This type of violence operates outside law and revolves around the lives of the subaltern. He writes,

> The tradition of the oppressed teaches us that the 'state of emergency' in which we live is not the exception but the rule. We must attain to a conception of history that is in keeping with this insight. Then we shall clearly realize that it is our task to bring about a real state of emergency, and this will improve our position in the struggle against Fascism.[39]

The Benjaminian distinction between the two forms of violence is vital in a revolutionary situation such as the Syrian one. It allows for a rethinking of the notion of sovereignty and violence in a context of state terrorism internally, and oppressive international law externally. The Syrian legal system legitimizes the state of emergency, and criminalizes any opposition to it. International law does not recognize non-state actors or subaltern violence outside the state. In short, it delegitimizes Benjamin's revolutionary violence. Applying Benjamin's notion of revolutionary violence in Syria provides a framework to counter the state of exception outside the legalistic framework described above.

Since there is revolutionary potential outside the polity, Agamben examines the ways life is transformed in a state of emergency. He makes a distinction between *bios*, which is the political life that operates within the polity, and *zoe*, which is "the simple fact of living" outside the political.[40] Zoe is therefore the form of life excluded from the polis, because it does not have a political dimension. It is prediscursive and cannot have any qualified form. "Bare life," for Agamben, is the introduction of zoe into the polis. What makes this form of life interesting for Agamben is that it cannot be controlled or captured by the political. It is ungovernable, and as such undermines the power of the sovereign. This is not an invitation to return to that form of life, but rather to understand the potential that resides in it. In addition, the concept of bare life and zoe can help to examine the continuum between life and death in the Syrian context. In addition, Agamben's critique of the oversimplified way sovereignty and the state of emergency are defined by international law allowing an exploration of the zones of indeterminacy in Syrian history. The Agambenian perspective allows for an understanding of the state of emergency outside the limiting framework of international law and human rights. It reveals state violence and the ways the legal system is instrumentalized to produce enemies rather than protect citizens.

While Agamben's work on the state of exception mostly focuses on Europe, it also provides important conceptual tools to examine the Middle East. His unwillingness to examine the Global South has led many post-colonial scholars to criticize his work as Eurocentric.[41] The scholarship that engages with his ideas in the context of the Middle East focuses primarily on the politics of Western involvement in the region. These studies are about neocolonial politics and Western implementation of the state of emergency in the Middle East. Many scholars invoke Agamben in their work about the US wars in Afghanistan and Iraq,[42] or Israeli's racist policies toward Palestinians.[43] There is, however, an emergent literature about the Middle East and Syria that engages with Agamben's work.[44]

BARE LIFE AND POLITICAL OPPONENTS

Quickly after seizing power, the Baath party produced internal and external enemies to mobilize the population and maintain supremacy over state institutions. In the early days, most of these enemies were within the Baath and hence the purges quickly took place within the party. Once Baath leaders who constituted a threat to Assad's rule were thrown in jail, the focus shifted to the Muslim Brotherhood and Iraqi Baath members. Since the 2011 revolt, the contours of internal and external enemies became blurred and overnight a large segment of the Syrian population became the enemy and was reduced to bare life.

Since it seized power on March 8, 1963, the Baath party has instrumentalized the state of emergency to crush its political enemies. The main enemy during the initial period was within the military and the party, as well as among the allies that supported its coup. The historic partner of the Baath party, Akram Hourani, was isolated because he disapproved of the union with Egypt and the Baath's close relationship with the Nasserists. Even the Nasserists, who helped the Baathists overthrow the Separatist Officers, were purged. Gradually, non-Alawi officers were removed from strategic positions within the army and replaced by Alawi officers. Michel Aflaq and his followers were marginalized at the Baath Sixth National Congress.[45] Salim Hatoum, a Druze Major, led a failed coup in September 1966, and was quickly pushed away along with other Druze officers. Between 1963 and 1970, there were several rebellions in Damascus, Hama, and Aleppo in addition to intra-party conflicts, which led to purges of military officers who opposed Assad and Jadid. In 1966, Jadid and the leftist branch of the Baath took control, but were deposed in 1970. The fact that Assad and Jadid relied on Alawis and other minorities to seize power in 1963, and the subsequent appointment of Alawis and family members in strategic positions within the military, led to the marginalization of Sunnis from the military and other positions of power within the regime. Early on, Assad developed a tactical prowess in sectarian maneuvering that slowly made sectarian politics one of the pillars of the Syrian regime, and one of the strategies for political control.[46] With each purge, Assad depleted his social base and alienated new segments of the population who supported Nasserism, Aflaq, or Hourani.

After Assad's 1970 coup and the Muslim Brotherhood's revolt in the mid-1970s, Islamists became the regime's new target. The legal system was redesigned and weaponized primarily against them. The Syrian regime built a matrix of power to control state and society. Article 8 in the Constitution of 1973 explains, "[t]he leading party in the society and the state is the Socialist Arab Baath Party."[47] To maintain control over society and state, Assad built a triangle of power, namely, the military and security apparatuses, the Baath party with more than a million members, and an extensive bureaucracy of the state.[48] The initial confrontation with the Muslim Brotherhood erupted when the requirement for the president to be Muslim was briefly removed from the new constitution. This led to intense opposition, which forced the regime to back down and reinstitute the religion of the head of state as Muslim. As the confrontation between the regime and the Muslim Brothers intensified, Assad tried to split their party by building a close relationship with the Damascus branch, which was led by Essam Attar. The other two branches, namely, the ones based in Hama and Aleppo, were more radical and did not believe the regime would make concessions or could be reformed. They chose a military confrontation with it, which would be violent and last until 1982.

In 1980, the Syrian Communist Party (Political Bureau) headed by Riyad el-Turk, the Democratic Arab Socialist Union (led by Jamal al-Atassi), and several smaller parties created the National Democratic Rally (NDR), a secular and progressive front. The Rally called for the overthrow of Assad's regime and the transition toward a democratic society. State violence against the Muslim Brotherhood intensified, and residents who lived in the wrong neighborhoods were indiscriminately killed for their supposed support of the Islamists. Aleppo was the target of Syrian forces in 1980, when several quarters (see Figure 2.3) were put under siege for extended periods before being attacked for their alleged support of the "gangs." French scholar Raphaël Lefèvre notes,

> [b]y mid-March 1980, units of the Third Army Division entered the city, their commander, General Shafiq Fayadh, warning the townspeople that he was "prepared to kill a thousand men a day to rid the city of the vermin of the Muslim Brothers."[49]

The NDR formed an alliance with the Muslim Brotherhood to oppose state violence, but its leaders and many members were put in jail, while others fled.[50] The regime understood the NDR represented a real threat to the regime's legitimacy because it proposed a democratic, progressive, and secular program that many Syrians would have supported.

After the death of Hafez al-Assad in 2000, public intellectuals and opposition leaders issued a statement that requested more openness and true democratic reforms. The Damascus Spring, which was confined to intellectuals and political circles, organized public discussions about some of the most pressing issues facing society, such as democracy in

Syria, economic justice, and the role of youth in society.[51] The NDR, which was banned in 1980, was revived in 2000 and called for the end of the state of emergency, as well as the release of political prisoners.[52] It also began opening cultural clubs and forums in most large cities. Those involved in the movement "may not have shared ideological affinities, but were nevertheless committed to some form of political reform within the country."[53]

Many leaders involved in the Damascus Spring had been previously in prison for many years for their opposition to the Assad regime. The regime recognized the potential threat that such a democratic movement could trigger, and thus repressed it swiftly. "In an interview of 9 February in the Saudi daily *Ash-Sharq al-Awsat*, the president called the activists of the Damascus Spring witting or unwitting enemy agents."[54] The forums were closed, and leading figures of the movement were put in jail or threatened by the Mukhabrat.[55] After crushing the Islamist rebellion in 1982, the regime perceived secular groups as the main threat. Citing an Egyptian journalist who was present in Damascus in 2010, Raphaël Lefèvre notes,

> a closed seminar on secularism at Damascus University attended by no more than 100 people from the ranks of progressives and democrats was banned by the authorities, while two weeks [earlier] a conservative cleric was allowed to preach in Aleppo, in the north of the country, in a sermon attended by 6,000 people.[56]

With the terrorist attacks of September 11, 2001, the Syrian regime borrowed from the Western dominant discourse about terrorism and counter-terrorism. Many post-colonial countries, including Syria, were more than eager to embrace the new discourse about the global War on Terror. To reinforce their power, these regimes were perfectly willing to attune their legal and repressive apparatuses to the United States new discourse about the war on terror and terrorism. In Syria, many were thrown in jail for "membership in an association created to change the economic or social structure of the state or the fundamental fabric of society" through "terrorist means."[57] The Syrian legal code was now in phase with its Western counterpart.

Shortly after the attack, the USA outsourced torture and arbitrary detention to Arab countries, including Syria. The Syrian prison system became an important node and an integral part in the global war against terrorism.⁵⁸ In every instance, the Syrian regime banned political activities and closed political spaces, and stripped individuals of their political rights, reducing them to bare life. The security approach of the Mukhabrat did not abide by any legal code. Detention without trial, torture, and summary killings were commonplace throughout the reign of the Assad family.

Finally, with the Arab revolts in 2011, the focus of the regime shifted to criminalizing popular protests. When the state of emergency was lifted in 2011 due to popular pressure, the regime quickly replaced it with counterterrorism laws in 2012 and instrumentalized the legal system against protesters and political opponents. In the same vein, counterterrorism courts were weaponized against prominent human rights activists.⁵⁹

The Syrian regime resorted to repression and violence every time it faced a political crisis. Instead of proposing a peaceful resolution, it eliminated its political opponents and opted for more repression. The central demand of Syrians, who opposed it, regardless of their political affiliation, was always to put an end to the state of emergency. In every instance, the regime criminalized its political opponents and stripped them of their basic rights. Regardless of their political affinities, they were treated as terrorists with no civic rights. Agamben explains that bare life (zoe) is a necessary outside that constitutes political life (bios). The sovereign is the one that defines the inside and outside of polity. While Agamben's conceptualization is useful for understanding the state of emergency beyond a legalistic framework, it does not sufficiently address the specificities of regions located in the peripheries of the West. Political life (bios) in the Western sense (political parties, freedom of the press, independent judiciary system, free elections, etc.) is non-existent in a despotic country such as Syria.⁶⁰ Since 1963, the Syrian state has systematically suppressed political spaces whenever they emerged. As such, there is a need to rethink the Agambenian perspective outside a European context.

NECROPOLITICS AND THE SYRIAN REVOLT

Achille Mbembe, a post-colonial intellectual from Cameroon, argues that the state of exception first emerged in the colony, not the metropole. Unlike Agamben, he is interested in exploring the topography of violence in colonized spaces instead of Europe. While Agamben examines the origins of the state of exception within the European continent, Mbembe shows that it is imperative to study its deployment in the colonies. His work can help us explore the implications of the state of emergency in a context such as the Syrian one.

Mbembe explains that Nazi genocidal politics, for example, would not have been possible without the history of European violence in the colonies. In other words, Nazism originated, on the one hand, in European colonialism, and on the other, in the development of bureaucracy in Europe. The convergence of these two moments produced new technologies to serialize and bureaucratize death. Mbembe writes, following the intellectual Enzo Traverso,

> ...the gas chambers and ovens were the culmination of a long process of dehumanization and industrializing death, one of the original features of which was to integrate instrumental rationality with the productive and administrative rationality of the modern Western world (the factory, the bureaucracy, the prison, the army).[61]

In a post-colonial context, the guiding question for Mbembe is, "Under what practical conditions is the right to kill, to allow to live, or to expose to death exercised?"[62] He explores this question through a critique of Michel Foucault's notion of biopolitics. The French intellectual developed the concept of biopolitics to understand the new mechanism of power in Europe. "Biopolitics" refers to the rationalities required for the administration of a population through mechanisms that foster life. The main problem is that Foucault does not pay enough attention to "politics as the work of death." To analyze the politics of death, Mbembe suggests that "necropolitics" rather than "biopolitics" is a more adequate conceptual framework to study the power of the sovereign in the Global South. Simply put, necropolitics refers to the creation of death zones where human lives are destroyed without tangible consequences for the sovereign.

Mbembe revisits Foucault's concept of biopolitics to show that necropolitics rather than biopolitics is the "lived experience" of the subaltern in the Global South. Necropolitics is the dark side of biopower in the same way that the colony is the dark side of the metropole. The state of exception for Mbembe is the space where necropower operates by deploying its lethal power and making decisions about who can live and who must die.

Mbembe explains that slavery is one of the most important modern manifestations of terror. As such, the slave experiences triple loss: "loss of a 'home,' loss of rights over his or her body, and loss of political status. This triple loss is identical with absolute domination, natal alienation, and social death (expulsion from humanity altogether)."[63] He explores the ways in which necropolitics circulates in space and how it takes place. The spatialization of necropolitics is an important element to understanding how this type of power operates. In other words, necropolitics creates "death worlds" where the enemy is targeted and where necropower operates. The destructive power of necropolitics is deployed in these death worlds to subjugate populations "to conditions of life conferring upon them the status of living dead."[64]

Mbembe's theoretical framework allows for a better understanding of how the politics of death operates in a post-colonial context such as the Syrian one. The colonial project of Europe could be understood as the externalization of violence from the European continent and its dissemination in the colonies. Mbembe notes,

> To properly assess the efficacy of the colony as a formation of terror, we need to take a detour into the European imaginary itself as it relates to the critical issue of the domestication of war and the creation of a European juridical order (*Jus publicum Europaeum*).[65]

The process of democratization in Europe produced better living conditions for European workers. There are evidently many exceptions to this general trend including the violence which workers and subaltern groups are regularly subjected to when they threaten the interests of the state and elite classes allied to it. As a result, white workers in Europe had access to socio-economic benefits unavailable to their counterparts in the Global South. The social contract between European workers and elites led to higher labor costs within the continent. Consequently, cheap labor and the conditions to sustain it had to be created outside Europe. The transatlantic slave trade and colonization were the solution for cheaper labor. Europe gradually exported its violence to the peripheries and created an uneven map of terror and exploitation. As such, the study of violence in the periphery is vital to better understanding the state of exception.

Agamben's study of the state of exception in Nazi Germany is therefore the second phase of what had already taken place in the peripheries.

Aimé Césaire, the anti-colonial intellectual from Martinique, writes about French colonial violence in *Discourse on Colonialism* (1972),

> First we must study how colonization works to *decivilize* the colonizer, to *brutalize* him in the true sense of the word, to degrade him, to awaken him to buried instincts, to covetousness, violence, race hatred, and moral relativism; and we must show that each time a head is cut off or an eye put out in Vietnam and in France they accept the fact, each time a little girl is raped and in France they accept the fact, each time a Madagascan is tortured and in France they accept the fact, civilization acquires another dead weight, a universal regression takes place, a gangrene sets in, a center of infection begins to spread; and that at the end of all these treaties that have been violated, all these lies that have been propagated, all these punitive expeditions that have been tolerated, all these prisoners who have been tied up and interrogated, all these patriots who have been tortured, at the end of all the racial pride that has been encouraged, all the boastfulness that has been displayed, a poison has been instilled into the veins of Europe and, slowly but surely, the continent proceeds toward *savagery*.
>
> And then one fine day the bourgeoisie is awakened by a terrific reverse shock: the gestapos are busy, the prisons fill up, the torturers around the racks invent, refine, discuss.[66]

Césaire emphasizes the weight of colonial violence in the production of savagery in Europe. By focusing on the state of exception in Europe without highlighting its origins in the colony, Agamben downplays the organic relationship between center and periphery.

While Mbembe's "necropolitics" addresses Agamben's Eurocentric framework, it brings a different tension when deployed in the Syrian context. The main issue with Mbembe's theoretical framework is that his topography of death revolves around the lived experience of the slave. The slave provides free labor, and as such must be kept alive. He "is therefore kept alive but in a *state of injury*."[67] Syrians, who are the objects of necropower, are not perceived as essential for the society or the economy. The primary object of the Syrian economy, as we will see in Chapter 4, is to help maintain power. Unlike the productive slave, whose life should be preserved since it is vital for the plantation economy, a Syrian citizen is not essential to the regime, and as such can be disposed of. In the Syrian context, necropolitical spaces are populated with people with no tangible economic value and are not worth keeping alive, especially when they threaten the despotic order.

The following section examines six aspects of necropolitical processes in Syria since 2011. These processes are not new for the most part, but they have been deployed more extensively since 2011. The prison system and dentition centers, which are central in necropolitics are discussed below. The codification of necropolitics in the Syrian context will require additional research, but the technologies of violence described below can help navigate the geography of violence in the current complex conflict.

Spectrums of death

A spectrum of violence was operating in Syria well before the revolution, but since 2011, it has been reconfigured. It allows the Syrian regime to deploy gradual techniques that create uneven death worlds. The spectrum of violence starts with the fear of being arbitrarily arrested and subjugated to torture. It includes the siege and subsequent politics of starvation. It involves the various ways Syrians are tortured and indiscriminately killed. In many of these cases, torture is not performed to gain information, but rather to actualize state power. The combination of direct and indirect violence that Syrians have experienced since 2011 has led to catastrophic humanitarian conditions.[68]

The Syrian regime and its allies have created a spectrum of death worlds where bodies are subjected to various forms of violence. The cruelty of a technique does not have a universal impact; its effect varies from one body to another. For example, crossing a checkpoint has an uneven impact, depending on various factors including the place of birth or the last name of a person. Some people have reported that their unusual last names have helped them bypass certain checkpoints, whereas others have disappeared or were killed because they had the wrong family names.[69] Since 2011, Syrians living in regime-controlled areas have been under a constant threat of being stopped, searched, and arrested. In many cases, residents of a neighborhood construct a mental map of the level of dangerousness in the surrounding areas. They know where the most dangerous checkpoints are and which ones are more lenient. For men, there is the fear of being conscripted into the army. Women are often sexually harassed and raped at these checkpoints. A UN Human Rights Council report details the violence women endure at checkpoints by government soldiers and pro-regime militias. In one case, a young woman who had been stopped at a checkpoint in a suburb of Damascus in October 2012 was taken to a military vehicle, subjected to mock executions, and raped by a Syrian army officer. Afterwards, the officer burned her hair and she was subsequently taken to a detention centre.[70]

In some cases, young men had to choose between two bad options. If they stayed home, they could face conscription; if they left, they might be killed by a sniper or arrested at a checkpoint.[71] Evidently, not all checkpoints are the same, as a 2014 study about Aleppo has shown.[72] The Karaj al-Hajez crossing point was known as the Death Checkpoint because many Aleppans lost their lives trying to cross it.[73] Then there are the various spatial strategies utilized by the regime to control the urban texture. They include roadblocks, snipers positioned on rooftops, or the siege of an undesirable area. The gradation in the politics of cruelty goes all the way to the starvation of entire quarters, such as al-Yarmouk camp in Damascus, as one example among many.[74]

The uneven deployment of the politics of death is used to punish and reward specific areas. For instance, the regime's killing machine can target a liberated neighborhood incrementally to break the will of its inhabitants. There is often strong correlation between a neighborhood or village's ability to develop successful grassroots politics and the level of punishment it receives. The more inhabitants are able to produce autonomous politics, the more they are perceived as a threat to sovereign power, and as a result, are punished.[75] In 2016, this was the case of Ma'art al-Nouman, which was well organized and was one of the first cities to be liberated in 2012. Four years later, the al-Nusra Front tried to take control of the city, but the population rose against it by organizing large, daily protests.[76] The regime perceived grassroots organizing against al-Nusra as a direct threat to its legitimacy. Through their struggles, protesters and organizers in Ma'art al-Nouman were essential in debunking the narrative that Assad had tried to impose since 2011. Since shortly after the first protests in 2011, Assad and state officials have insisted there is no authentic revolt in Syria, but only Sunni terrorist groups terrifying minorities.[77] A few days after the protests in Ma'art al-Nouman, the regime's jets targeted popular markets multiple times, killing dozens of civilians.[78]

The same logic of cruelty applied to individuals who played a crucial role in the revolt. For example, a media activist who videotaped a demonstration and uploaded it to YouTube would get a harsher treatment than a protester. Former prisoners explain that in many cases, media activists and doctors were subjected to more torture than FSA fighters, because they constituted a greater threat to the regime.[79]

Scales of death

A second lens through which one might examine the politics of death in Syria is the scale of violence. Necropolitical processes operate at different scales, starting at the micro-level of the body and going all the way to the macro-level that comprises an entire city. They can be deployed to inflict pain with precise intensity on specific parts of a prisoner's body, as well as to target an entire neighborhood with chemical weapons. In addition, these necropolitical processes work at the level of an entire nation and should be understood as part of a genocidal politics. One of the main implications of necropolitics on Syria is how the combined effect of various techniques initially curtailed, and later reversed the growth of the population. At the national level, the size of the Syrian population fell from 21 million in 2010 to 17.7 million at the end of 2014. The combination of displacement and deaths led to a 23 percent drop in the projected size of the population in 2014.[80]

If a region is too difficult or costly to control, due to the inhabitants' organizing and level of resistance, the regime considers it to be enemy territory and targets it indiscriminately. This was the case of Dara'a in the south, Homs in the west, Idlib province in the northwest, and Hassakah in the northeast. Other areas were more docile in the beginning because of the violence they experienced in the past. Aleppo and Hama had witnessed the violence of the regime prior to the 2011 revolt.[81] This could, in part, explain why certain social groups in Aleppo were initially reluctant to participate in large numbers in the protests.[82] In addition, the regime built an extensive network of surveillance and repression in certain areas, making it risky to protest. It used the Berri Clan in Aleppo to repress the revolt and terrorize the population.[83] This is, in part, why the level and intensity of protests were uneven in different regions. It is worth noting that the regime's geography of violence correlates closely with its fear from specific areas. The more organized protesters in a city or a village are, the more brutal the militia's or security branch's response.

The distribution of violence follows a precise sectarian logic. The regime used sectarian arguments to justify its large-scale campaigns against Sunni areas that opposed it. Opposition coming from non-Sunni regions required micro-scale responses. In other words, the Sunni regions could be collectively punished, whereas such an action would be risky outside these regions. For instance, Latakia, a coastal city with a large Alawi population, witnessed a number of protests in 2011. The regime used snipers to repress protesters and blamed "armed gangs" for targeting residents from rooftops.[84] Suweida, a city populated mostly by Druze, had several waves of protests and a massive refusal to join Assad's army. The Sheikhs of Dignity and the prominent Sheikh Wahid al-Balous supported the protestors. To avoid a full-scale confrontation with the Druze community, the regime opted for micro-scale strategies. It began with the assassination of political opponents, including al-Balous.[85] Then the regime created a Druze front to counter the rebellion from within, instead of using its Alawite and Shia militias, who might have sparked sectarian strife against Assad.[86] Finally, it made it easier for ISIS to attack Suweida, to remind Druze that any alternative is worse than the regime.[87]

The same strategy of assassination and repression through local militias was employed in the Kurdish regions in the north. Mashaal Tammo, a progressive intellectual and a grassroots leader in the Kurdish region, was assassinated in his home, most likely by pro-regime gunmen.[88] The Syrian regime, with the tacit help of the Democratic Union Party (PYD) repressed Kurdish grassroots rebellion and as such avoided a confrontation between Arab militias and the residents of these regions, many of whom are Kurdish.[89] The regime's politics of death deployed in Druze and Kurdish regions is about utilizing fine-tuned techniques rather than collective punishment, which was commonplace in most Sunni opposition areas.

The regime's fine-tuning in non-Sunni regions was supplemented with an openly sectarian logic that mobilized some segments of minority groups against a rebellion that Assad presented as sectarian.[90] This sectarian logic allowed for the use of large-scale destruction in Sunni regions, and the deployment of minimalist strategies in non-Sunni areas. In the end, necropolitics operates in different ways to create uneven territories where the scale of violence is determined according to the ways the regime defines the enemy and its territories.

In 2013, the Syrian army and allied militias were overstretched and unable to sustain confrontations on many fronts. Assad explained in a speech that it was essential to maintain control over what he defined as the "Useful Syria."[91] That region includes the most populated areas, the strategic coastal region in addition to Damascus and Aleppo, and the axis that connects them to each other. Anything outside that region was deemed un-useful and unnecessary for the survival of the regime. It was a region where the killing machine of the regime could operate on a large scale and with high intensity, without restrains or constrains. What is revealing about "Useful Syria" is what lies outside it or what the regime considered "Useless," and as such could be disposed of. Useless Syria is the equivalent of Agamben's camp, which is the space located outside the polity. It signals that lives (or zoe) in these geographic or conceptual spaces are worthless and can be obliterated without any real repercussions.

Velocity of death

To preserve sovereignty, the state deploys multiple death worlds operating at different speeds. For example, snipers positioned on rooftops can control several strategic axes in a city. The sniper takes away life instantly, if a person crosses a prohibited line. Evidently, the rebels have used snipers to target the regime's soldiers; however, due to its domination of the sky, the regime has near absolute control over high buildings and structures where snipers are positioned. Aleppo is one such example where snipers control the tempo of death in the city. The Syrian state and its Russian allies have absolute control over vertical power. Their air forces can hit any target anywhere in Syria and cause immediate death. Further, the regime has various tools at its disposal to inflict slow death on entire regions. While the opposition did impose a criminal siege for a short period on Western Aleppo, and longer ones on Fuaa and Kafraya in northwestern Syria, the Syrian government has conducted most sieges.[92] Slow death usually requires fewer resources and is less costly to the imposer; this is why the regime uses it on a large scale. In some cases, slow death is used because the regime's military forces are overstretched and cannot fight on more fronts—either because they lack manpower or military equipment. In these cases, and when possible, the regime imposes a siege with devastating implications.

The medical field is one of the preferred targets of necropower. When medical facilities are targeted, several temporalities are often at work. By preventing medications and doctors from entering a besieged area, the regime effectively causes the slow death of the inhabitants. Mbembe explains that necropolitics is the "subjugation of life to the power of death." The multiple velocities of death are used to manage the deterioration of an entire population. In certain cases and when the regime is not facing an eminent threat, the siege and the slow death that results is preferable to the immediate death due to barrel bombs.

The Syrian American Medical Society explains that in Ghouta, which is located in Eastern Damascus, "Many are simply awaiting their death at home."[93] Their report, titled "Slow Death: Life and Death in Syrian Communities Under Siege," documents the operationalization of necropolitics in these areas.[94] Even the United Nations has played an important role in covering up for the regime and helping it weaponize the medical field against civilians.[95] The UN has delivered all humanitarian aid to the Syrian government, ignoring many NGOs' pleas not to do so, and as a consequence, making it very difficult, if not impossible, for such aid to reach opposition areas.[96] On the other end of the velocity spectrum, the regime's jets have targeted hospitals and ambulances, causing instant death while destroying the medical infrastructures in many cities.[97] The destruction of medical facilities is particularly cruel in wartime, but was nonetheless very efficient in breaking the cities in the liberated regions.

The prison system is another institution where different velocities of death are working together. The slow process of wearing out prisoners is part of a long-standing strategy in Syria. It can be accelerated or slowed down through various technologies of death. Jalal Mando, a media activist from Homs, witnessed during his arbitrary detention the sadistic torture and summary executions of many prisoners. In the Palestine Branch, a renowned security facility, he survived what he calls the Diarrhea Massacre. On New Year's Eve 2014, a guard put laxative in the food without informing the prisoners; due to their frail bodies and weak immune systems, 35 inmates died that night.

Jalal explains that every prison is required to deliver, on a weekly basis, a specific number of corpses. If on a given week the Branch does not meet the required number of dead prisoners, then some individuals are selected to receive an air injection in their arterial lines and die quickly.[98] In every cell, there is a *homo sacer*, a prisoner who is on his way to death due to starvation and exhaustion. He disconnects from his surroundings and stops eating or talking, gradually moving into a death zone or a liminal state. Neither dead nor alive.[99] The *homo sacer*, or the disconnected, is usually the one chosen by the guards to kill to meet the threshold number.

Slow and quick deaths are often used simultaneously to achieve greater destruction. Throughout the war, the regime deployed slow death in certain areas as a form of collective punishment, while at the same time using its military capabilities elsewhere to achieve quicker results. These different temporalities of death were used effectively by the regime to crush the revolt and gradually subdue the population.

Remoteness of death

Death at close proximity usually means that its occurrence does not entail mediation. It is a death that involves intense manual work. A prisoner recounts that during torture sessions, the prison guards who did not beat an inmate relentlessly for hours without breaks could face torture themselves.[100] A death at a distance requires the deployment of adequate technologies: the bullet of a sniper, a barrel bomb dropped from a helicopter, or a rocket launched from a nearby base.

The target of the war machine can be in the vicinity or not. Individuals can be killed with long-range weapons fired hundreds of miles away. On the opposite side of the spectrum, a prisoner can be killed in close proximity and by the bare hands of his torturer.

Necropolitics does not necessarily stop after the death of a prisoner. There are countless accounts of bodies that have been disfigured or dismembered after their deaths. Former prisoners have also provided detailed stories of large-scale operations of organ harvesting in Syrian prisons and hospitals.[101] In the latter case, necropolitics works from within the body: it involves the collection of an organ that could potentially save the life of a militia fighter, and as a result revitalizes the killing machine.

The dangerousness of an individual determines whether he or she needs to be detained and tortured. Activists and opposition fighters experience the cruelest treatment, and are often tortured to death if arrested at a checkpoint. They frequently die in proximity of their torturer in the interrogation chamber. Many pro-regime doctors work in prisons to monitor the prisoner's tortured body and assess whether it has reached its limit or can withstand additional pain. These death doctors use their expertise to guide the guard during the interrogation, to bring the body as close to death as possible but still prevent its demise.[102] The collaboration between the prison system and the medical field produced new death worlds where the limits of the possible were extended. Those who survived their stay in Hospital 601 in Damascus tell nauseating stories about the treatment of prisoners, and how many preferred to go back to the horrifying prison rather than stay in the "slaughterhouse."[103]

Those who are killed at a distance by the war machine usually inhabit a space the regime considers as threatening. They die as part of the collective punishment in enemy territory. Government militias were behind several sectarian massacres in the first two years of the conflict. The main purpose was to scare religious minorities and deter them from joining the rebellion. These massacres required direct contact with the body of the victim, and their outcomes were gruesome. They were used to dissuade people from participating in protests and to turn a popular revolt into a sectarian conflict.[104]

Accuracy of death

This dimension overlaps with several others described above. Oftentimes, closeness leads to more precise targeting, but this is not always the case. The sniper, for example, can kill his target with high precision from a distance. Necropolitics can be deployed to target bodies with high precision, but it can also be used in heuristic fashion against unruly territories. It punishes entire groups who live in the wrong neighborhood. The targeting of civilian sites with high precision can be very effective in terrorizing inhabitants and pushing them to surrender. Mbembe notes how accuracy and purposeful *in*accuracy are often two facets of the same necropolitical process. He notes,

> [o]ccupation of the skies [...] acquires a critical importance, since most of the policing is done from the air. Various other technologies are mobilized to this effect [...] Killing becomes precisely targeted. Such precision is combined with the tactics of medieval siege warfare adapted to the networked sprawl of urban refugee camps.[105]

To take one example, the Syrian regime targeted medical facilities in Aleppo with high precision. Between June and December 2016, the Syrian army and its allies completely destroyed medical facilities in the city by targeting them 73 times, in most cases multiple times each.[106] The combination of precise attacks on vital facilities such as bakeries, gas stations, and hospitals, as well as the indiscriminate and massive attacks on east Aleppo with inaccurate barrel bombs, led to the fall of the city in December 2016.[107]

(In)Visibility of death

The state of emergency creates insides and outsides as well as zones of indeterminacy where the legal is suspended. Death worlds operate in uneven ways depending on the exposure they get and the type of witnesses that inhabit them. Torture inside a prison is invisible to those outside the facility, but is unabashedly visible to those inside. When ISIS tortures its prisoners to death and films them with high-quality equipment for the world to see, its goal is to create a death world where the inside and outside are indistinguishable. The difference between ISIS and the Syrian state is not about the level of cruelty, but rather the degree of visibility. When pro-regime militias kill more than 100 Aleppans and throw them in the river, knowing that the water will drag their bodies to east Aleppo, which is controlled by the opposition, their purpose is that the inhabitants of these neighborhoods see the massacre.[108] The violence of such an act is mostly invisible to the outside, especially when media and images are easily manipulated, but is unmistakably visible to the inside. The question of visibility is essential, since it is used by the state to remind the population of the high cost of opposing it. On the one hand, visibility of violence allows the regime to re-educate the rebellious population; on the other, it hides the violence from those who do not need re-education, since they adhere to the same hegemonic framework. In addition, the regime has an interest in hiding the violence from powerful foreign players who could use graphic images to undermine its legitimacy. In some cases, inside and outside are not clearly defined. Agamben notes that:

> the state of exception is neither external nor internal to the juridical order, and the problem of defining it concerns precisely a threshold, or a zone of indifference, where inside and outside do not exclude each other but rather blur with each other.[109]

In August 2012, the Syrian army massacred more than 500 inhabitants in Daraya, southwest of Damascus, and left "Assad or we burn the country" inscriptions on the walls. Shortly after the tragedy, a pro-regime journalist embedded in the Syrian army was dispatched to the crime scene to ask children holding their dead mother indecent questions.[110] Surrounded by soldiers who had massacred the inhabitants, the same journalist interviewed an injured woman languishing in a cemetery instead of rescuing her.¹¹¹ In such instances, the boundaries between inside and outside are blurred. The regime's barbarism is shown to terrify the enemy and mislead their supporters. Overall, the combination of these techniques of state terror should be understood as part of a genocidal politics.¹¹²

This process of reorganizing society in a way that makes genocide acceptable to a majority is essential. This process is gradual, as the allies progressively internalize the regime's narrative about the "enemy." For many, it starts with a complicit silence, which gradually becomes a vocal approbation. Throughout the process, the production of the enemy, namely, as a foreign agent, a Takfiri Salafist, or a terrorist, is an important step toward imposing a politics of death that is acceptable to a large segment of the population.¹¹³

THE PRISON AS A CASE STUDY

This section presents a brief history of the prison system in Syria and then proposes a conceptual framework to analyze its current dynamics. The prison system played a central role during the Syrian revolt, and is the central pillar of necropower. According to the Syrian Network for Human Rights, the Syrian government arbitrarily detains almost 90 percent of all prisoners in Syria as of 2019. The remaining 10 percent are in ISIS, PYD, and opposition groups' detention centers.¹¹⁴ Syrian writer and former political prisoner Mustafa Khalifeh explains, "[t]he history of Syria is a history of prison, concentration camps and massacres."¹¹⁵ Prisons should be viewed as one strategy of violence among several others. As discussed above, there is a continuum of violence in Syria, and the prison is only one tool among many.

The modern Syrian carceral system has a long history that begins with the French colonial occupation in the 1920s. The French mandate relied on a complex network of surveillance and coercion to maintain power and prevent revolts against the colonial order. Revolutionaries were severely punished, and in the event they presented an imminent danger, were publicly executed. During the French mandate, prisoners were often used as forced labor.¹¹⁶ After independence, Husni al-Zaim and Adib al-Shishakli, who seized power through military coups respectively in 1949 and 1951, both expanded the police state and the prison system.

While the Syrian regime utilizes a wide array of strategies to maintain power, the carceral system is a central institution and essential for its existence. The Syrian prison system is the necropolitical space *par excellence*, and it is the institution where the state of exception operates seamlessly, with little obstruction. Political prisoners spend many years, oftentimes several decades, in prison without trial or communication with the outside world. Torture is part of their daily routine, while humiliation and shaming is an essential part of the prison system. Syrian novelist Ibrahim Samuel, who spent many years in prison and wrote about his experience afterwards, explains, "[m]y challenge to myself was to convey the absolute simultaneity of life and death that is part of the prison experience."[117] What he describes is the zone of death, where life and death become indistinguishable.

Sednaya prisoners depict a horrific picture of inmates who cannot endure more torture and therefore disconnect from their surrounding by refusing to eat or speak. They slowly walk toward their demise.[118] American academic Miriam Cooke, who has written extensively about dictatorship and prisons in Syria, notes that "[p]rison narratives provide a prism onto life under authoritarian rule."[119] The politics of death under Hafez al-Assad reached a peak in the 1980s. Most of the violence during that period was deployed within the massive and secretive prison system that the Syrian regime had built; it was also, in part, inherited from the French colonial power and the autocratic leaders who ruled Syria after independence.

Any effort to understand necropolitical processes in Syria requires a thorough examination of the carceral system. Several things have made this institution particularly important in Syria since 1963. First, the regime uses it extensively to punish its enemies and dissuade Syrians from opposing its rule. With the exception of a few short periods of open defiance to the Assad regime, prison as an institution was effective in silencing dissent and producing a docile population. Second, prison provides a space to conceal state violence from the larger society. The level of violence deployed in Tadmur and various security branches was mostly ignored by a large segment of Syrian society until 2011. Third, prison is used to incarcerate the unproductive surplus population. With high unemployment rates and a section of the poor classes involved in petty crime, prison performs its basic task of locking up the unruly reserve army of labor. However, the distinction between political and criminal prisoners is often problematic, since it is the ruling classes that create this taxonomy. Laleh Khalili and Jillian Schwedler note,

> [D]rawing a clear boundary between political and criminal prisoners is problematic. Criminalizing armed struggle in a counterinsurgency, for example, would increase the number of "criminal" prisoners, where in fact, those prisoners would consider themselves political. Beyond such conceptual confusion, "crimes" itself is a political category, over-determined by the constantly changing mores and norms [...] and by the definition of politics itself.[120]

The Syrian revolt has shown that the boundaries between the political and criminal are at best problematic, and at worst part of the necropolitical process.

Riyad al-Turk, the leftist opposition leader and former secretary-general of the Communist Party (Political Bureau), spent close to 20 years in prison under Hafez al-Assad and his son Bashar. He dubbed Syria the "Kingdom of Silence," because no criticism of the regime, no matter how mild, was permissible. The mass production of silence in Syrian society revolves around a violent police state, surveillance agencies, and a robust prison system. Citing Frank Graziano, Sune Haugbolle notes,

> Rumours of violence renders the public both "audience" because it "witnesses" the abstract spectacle of detention centers [and] "actors" because its status as audience—however passive it may appear—is a function integral to the efficacy of the spectacle by which power is being generated.[121]

The Syrian carceral state uses a vast array of techniques to impose this silence. If self-censorship and intimidation do not work, the regime deploys the various technologies of death at its disposal, and the prison system is its cornerstone.

Mustafa Khalifeh, who spent 13 years in ten different prisons (including the infamous Tadmur prison), explains, "[t]he regime wanted to empty society from politics."[122] He points out, "Syria's history is a history of prisons, concentration camps, and massacres." Without a thorough understanding of the prison system, it is impossible to comprehend the mechanisms of power in Syria. Khalifa asserts the centrality of prison in the Syrian topology of violence. He explains, "this regime came through violence, it has maintained power through violence, and it won't be eradicated without violence."[123]

It is essential not to view the prison system in Syria as simply a space of incarceration where the technologies of death are deployed against prisoners. This binary creates two distinct spaces (inside and outside) that are clearly separated. According to this view, necropolitics operates within well-defined spaces, but would typically be non-existent outside them. This minimalist definition that views prison as the space of incarceration prevents a full understanding of how the prison system operates in the Syrian context. At the same time, it is essential to avoid the maximalist view, according to which under Assad's rule, the entire Syrian society lives in a large prison. Yassin Haj Saleh warns against such a maximalist definition, where the specificity of the prison experience disappears or is conflated with life experience outside it. As he notes, an expansive definition where distinctions between inside and outside are eliminated would be useless and ineffective to comprehend the strategies of power in Syria.[124] At the same time, it is vital to highlight the shortcomings of the minimalist approach and its restrictive boundaries. The Syrian revolt shows that necropolitical processes operate well beyond the narrowly defined boundaries of prison.

The Syrian prison system should be understood as a complex institution, not simply the built space used to incarcerate inmates. Unlike prisons in the West, its primary purpose is to lock up political prisoners. While the main function of prisons in the United States is to maintain a racial order and generate profit,[125] prisons in Syria operate at an economic loss, and their primary purpose is to maintain the Assad regime in power. The US prison became a central institution for the policing of the Black body and crushing resistance of the Civil Rights Movement. The "prison-industrial complex" plays a vital role in oppressing racial minorities and removing the disruptive surplus labor from public arenas. In addition, US politicians and companies have turned the coercive institution into a lucrative business.[126] Yassin Haj Saleh explains that prisons in Syria are unproductive in the economic sense, and their management is not bureaucratized the way it is in Western countries.[127] It is precisely because the human body is unproductive in Syria that it is disposable and can be tortured. Since the prison system in Syria defies any economic rationale, it could be viewed as a "prison *non*-industrial complex."

While there is no economic incentive for the Syrian prison system, unlike its US counterpart, both are complex institutions. The US prison industrial complex is composed of several institutions, including the media, construction companies, lobbies and politicians, prison guard unions, etc. Likewise, Syrian prisons should be understood as a complex network of various institutions, and as such can not be reduced to just the space of incarceration. The Syrian revolt has shown that strict separation between the checkpoints, the patrolling vans, and the network of snipers on the one hand, and the prison system on the other, is not sustainable. The prison system is a complex network with multiple nodes that include incarceration centers, security branches, checkpoints, and militias patrolling at night. The Syrian regime uses a multidimensional set of violent techniques, including intimidation and threats against individuals and their families as well as political assassinations. Snipers constitute a central node in the complex carceral network. They regulate the space through their accurately calculated positioning. They shoot enemy targets in opposition neighborhoods, but also provide vertical control over regime areas through surveillance networks by, for example, preventing residents from circumventing a checkpoint through back alleys. During the revolt, the carceral system was expanded significantly by turning new spaces, including schools, Baath party offices, hospitals, and official buildings into detention centers.

One way to examine the scope of the prison system is through Agamben. The Italian philosopher explains that the killing machine needs to identify a specific space to operate. That space, according to Agamben, is not necessarily the one the enemy occupies. For him, the distinction between friend and enemy is not a valid one in the modern era. In a state of emergency, argues Agamben, power does not make such a distinction, but is rather interested in the figures identified above, namely, zoe and bios. In such a context, there is "absolute indeterminacy between inside and outside." Likewise, it is impossible to distinguish between "observance and transgression of the law."[128] Following Agamben, there is a need to carve a space in between the minimalist conceptualization that confines prison to the enclosed space and the maximalist one that abolishes all boundaries.

CONCLUSION

This chapter examined the geography of violence in Syria since the Baath rise to power in 1963. One of the most effective tools the regime used against its political opponents is the state of emergency. It was utilized as an anti-politics machine to prevent the emergence of political spaces. When the Syrian revolt broke out in 2011, protesters' most important demand was an end to the state of emergency and the release of political prisoners. During the uprising, the regime deployed an array of technologies of violence to crush the protests. These techniques were used effectively to undermine grassroots struggles. Today, the prison system is producing death at an industrial scale. Its boundaries have expanded to include new spaces while its scope is virtually unlimited.

2
The Geography of Death in Aleppo

Leave, convert, or die.
    King Ferdinand and Queen Isabella to the Jews of Spain

Down with the regime and the opposition...
Down with the Arab and Muslim Worlds...
Down with the United Nations Security Council...
Down with the World...
Down with Everything...
    Occupied Kafranbel, October 14, 2011

Aleppo, once Syria's capital of classical music, sophisticated cuisine, and Islamic culture, today lies in ruins. Insurgents controlled two-thirds of the city for four years before it fell to the regime in December 2016. This chapter begins with a brief urban history that explores the ways Aleppo's urban fabric has evolved since the mid-nineteenth century. It argues that the city's urban forms from various historical stages, including the Ottoman Empire, French Mandate, and post-independence, have been reorganized and utilized by the Syrian forces since 2011 to break the city. Using the concept of "urbicide" (the deliberate and systematic destruction of a city), the chapter investigates the various spatial techniques developed by the Syrian regime to destroy Aleppo and its people. These techniques include the strategic positioning of snipers on tall buildings in a vast network throughout the city to segment space and block the circulation of civilians and insurgents. Snipers produced a cartography of fear that terrorized Aleppo residents, altered the tempo of life and death, and killed thousands. In addition, the Syrian regime weaponized the demographics of the city to crush the revolt. The Syrian forces instrumentalized the ethnic, religious, cultural, and class composition of the city to fracture the urban fabric and prevent the crystallization of revolutionary identities or urban solidarities. Finally, urbicide in Aleppo took the form of siege, destruction of infrastructure, and displacement of the population. As a result, east Aleppo was emptied of its population; epidemics and starvation were rampant; the death toll was staggering; and life expectancy dropped by more than a decade.

OTTOMAN ALEPPO

Aleppo was the third largest city in the Ottoman Empire and a main economic hub between the sixteenth and mid-nineteenth centuries. Merchants traveled long distances to buy Aleppo's luxury goods.[1] European countries maintained consuls in the city from the sixteenth century onwards to trade with Muslim merchants. By the mid-eighteenth century, Europeans gradually began to establish commercial relationships with Christians, which weakened the influence of Sunni notables in the city.[2]

To modernize its institutions and address structural problems, the Ottomans introduced the Tanzimat in 1839, which were major reforms to help the Empire better integrate the world economy, control nationalist movements within its territories, and to curb the increasing influence of European countries outside it. These reforms led to significant transformations in Aleppo, including a new taxation system, land reform, and a recruitment campaign for a modern army. The Tanzimat were not well received by the local bureaucrats, who believed Istanbul was using them to impose direct control over the provinces. The Muslim population was frustrated with the new tax, which until then was imposed on non-Muslims only. In addition, the construction of new churches was not well received among the same Muslim population. Ross Burns notes,

> As Aleppo became increasingly drawn into the world economy, [...] the Christians [...] could easily slip into roles as the point of contact with outside importers or exporters. The barriers to full Christian participation in society were being torn down by the Ottoman administration [...].[3]

Prior to the 1850 riots, there were frequent confrontations between Janissaries and Ashrafs. The former were the Ottoman Empire's elite infantry, but were disbanded in 1826 due to their refusal to follow Sultan Mahmud II's orders. They mostly lived in eastern Aleppo and had a strong corporatist identity. The Ashraf, the descendants of the Prophet, were the city's elite and were based in western Aleppo.[4] The residents of eastern Aleppo did not play any significant role in A'yan politics (notable politics) and were politically marginalized.

The combination of these factors led to a riot in east Aleppo on October 17, 1850. The Janissaries, Kurds, Turkmen, and Bedouin who were economically and politically marginalized and lived in eastern Aleppo attacked the Christian northern districts, namely, Jdeideh and Salibeh. The rioters looted the wealthy Christians' stores in the bazaar, attacked churches and individual Christian homes, and killed several dozen people.[5]

After the riot, the British put pressure on the Ottoman government and requested that Muslim residents repay an indemnity to Christian merchants and return their properties. The wealthy Aleppans were humiliated and refused to pay the indemnity, as it would have implied "that they were collectively guilty for something that 'Bedouins and Kurds' had done."[6] A second round of fighting between eastern and western neighborhoods erupted in November 1850, when the Ottoman central government sent its army to crush the rebellion. In the end, the army killed thousands in the eastern neighborhoods of Bab al-Nayrab, Qarliq, and Quarliq.[7] It is worth noting that the revolts of Damascus and Mount Lebanon in 1860 did not spread to Aleppo, primarily due to the fear of retaliation, in the event its inhabitants protested.[8] Ottoman violence in Aleppo a decade earlier had a lasting impact and residents were reluctant to participate out of fear of retaliation.

The city witnessed a major urban transformation at the end of the nineteenth century as a result of the Tanzimat. New streets and squares were built, as well as a tramway and other infrastructures.[9] Wealthy Aleppans moved to the west into newly established Christian neighborhoods such as Aziziyya and Hamidiyya, as well as Jamiliyya, a Muslim and Jewish quarter. In the second half of the nineteenth century, these new neighborhoods had all the amenities of modern life, including access to electricity and fresh water (see Figure 2.1).[10] Historian Nora Lafi notes, "[these neighborhoods] embodied the imperial modernity. Engineers and planners, including foreign experts, were asked to connect this new city to the old one, in a logic that was more one of integration than of juxtaposition."[11] The social and urban reorganization reinforced the split between west and east and still exists in present-day Aleppo.

THE FRENCH MANDATE

After the Ottoman Empire's partitioning, the French colonial power seized Syria and Lebanon, despite the local population's resistance. The nationalist revolts brought together a wide array of social and political groups who fought the French from 1918 to 1927. After World War I and the dismantling of the Ottoman Empire, Aleppo's economy began declining, primarily because it was cut off from its hinterland. The new borders between Iraq and Syria made it more difficult for the city to trade with Iraqi cities, which were essential to its economy.[12] The structural transformations of the economy and politics in Aleppo had major implications on the social composition of urban classes. Historian Keith David Watenpaugh explains that it is vital to avoid reductive binaries that, in the case of Aleppo, oppose the wealthy center to the periphery. He explains that the middle classes, the educated professionals, opposed traditional elites in Aleppo. They were not necessarily a social class in the economic sense. Instead, they built an identity in opposition to the Aleppine notables and the marginalized groups living in the east. The middle class was able to carve a space for itself in Aleppo's politics through its mastery of what Watenpaugh calls the technologies of the public sphere, which include the production of public discourse and the utilization of mass media to channel their ideas about nationalism, secularism, and modernity. After World War I and the French occupation of Syria, the middle class tried to negotiate a new position in which the Ottoman Empire would still play a role, while at the same time producing a new identity that was both Syrian and Arab. Aleppo's political elite was divided between those who sided with the Arab government led by Emir Faysal, who was based in Damascus, and those who preferred a form of autonomy with southwestern Anatolia, since it remained an important market for Aleppo's merchandise.[13]

The French attempted to reorganize the urban structure of Aleppo to undermine the resistance and counter anti-French sentiment. In 1929, the French Mandate asked urbanist Danger brothers to produce a master plan for Aleppo. Their first plan proposed to separate the Old City from the new western section, and destroy segments of the historic zone. The design reproduced a colonial trope according to which the Ottoman city is irrational and anomic, and consequently must be quarantined from the modern segment. Describing Aleppo in 1932, René Danger wrote,

[t]he streets are mysterious, with capricious twists [...] They evoke the Middle Ages and wars of religion, treachery and revenge, battles on doorsteps and in dead ends. Once the threshold has been crossed, the courtyard provides a contrasting image of a quiet and pleasant life in the cool privacy of open courtyards around a reflecting pool of water.[14]

The colonial power rejected the plan out of fear it would be opposed in the League of Nations.[15] The Danger brothers proposed a second plan in which the old Arab city would be preserved and modern urban spaces would be built around it. The contrast between the two was stark. French Orientalist Jean Sauvaget contrasted the regularity and order of the antique Roman city to the irregularity of the Muslim city. For him, "The Aleppo of the Ottomans is nothing but an illusion (*'un trompe l'oeil'*)—a sumptuous facade behind which there are only ruins."[16]

Daniel Neep explains that the colonial power in Syria deployed an array of techniques to manage the Syrian population and geography. In his book *Occupying Syria under the French Mandate* (2012), he explores the various forms of colonial violence the French used to crush the rebellion in the 1920s and 1930s. Using a Foucaudian framework, he analyzes several forms of power French administrators and military experts used to reorganize the urban geography of Damascus and Aleppo. Based on Orientalist knowledge about an "unchanging" and "primitive" Middle East, they developed a spatial strategy to quell resistance.

Neep explains that the French developed a twofold strategy to undermine a mounting opposition to their rule. The first part was based on disciplinary spaces that fragment and isolate the rebels that were based at different times in the Syrian North and South, as well as al-Ghouta. The colonial military and administrators found that classical models of warfare were ineffective in Syria. Instead of complete domination of a space, which is unpractical and difficult, they employed a strategy that required the compartmentalization of space to hamper the movement of rebels. These disciplinary spaces undermined the resistance of rebels and made their movements from one area to another much more challenging. The second technique was developed to address the problem of what they perceived as the primitive urban spaces in cities such as Aleppo or Damascus.[17]

Colonial administrators believed that urbanism could resolve many urban anomalies they perceived as inherent to the region. By doing so, they wanted to facilitate the French "civilizing" mission in Syria. Some of these colonial strategies were brought from other French colonies and deployed in the Syrian context. According to oriental knowledge developed in the colonies, urban planners believed the Old City should be preserved as an unchanging urban form and the native population in it should be maintained. The modern city should be created outside the perimeter of the old one. In between, a sanitary cordon should be inserted to make the separation permanent and prevent old and new spaces from intermixing. The Sunni Muslims would preserve their traditions in these impenetrable spaces while simultaneously undergoing a major "cleansing" that would ultimately improve their livelihood and keep them content, but also prevent the spreading of epidemics within the Old City, and, more importantly outside it, in the European section.

The European city would follow a rational urban matrix where residential and industrial zones would be separated and where access and flow would be prioritized for hygienic and military reasons. What is important for Neep is that the boundaries between civilian and military spaces are oftentimes blurred. The techniques utilized in urbanism often migrate and are redeployed for military purposes.

POST-INDEPENDENCE

In 1954, André Gutton, a professor at the École nationale supérieure des Beaux-Arts, proposed a new master plan for Aleppo. His modernist plan required the creation of new streets that would have fragmented the old Aleppo and destroyed important sections of the historic city. Gutton's guiding principles were: 1) to reorder the Old City; 2) to maintain better sanitation; and 3) to emphasize historic monuments. The Old City, according to him, needed to be cleansed and turned into a museum. By viewing the city as a labyrinth-like structure that required order, Gutton, like the colonial power before him, wanted to end the spirit of clans that resisted modernity and permeated the old neighborhoods.[18]

Overall, the Old City was simply neglected for several decades, since none of the plans were implemented, due to their flaws. Consequently, many urban notables began leaving the Old City and renting out their houses to poorer or rural families. Their preferred destination was west Aleppo, because it offered all the amenities of modern life. The Syrian government implemented the part of the Gutton plan that focused on car circulation and ignored the remaining aspects of the plan.

After the 1963 coup and the Baath party's rise to power, urban notables were weakened, in accordance to the new ideology, and their houses in the Old City were seized and turned into multiple smaller apartments. Jews in Aleppo were attacked and their houses burnt, while many were forced into exile.[19] Gyoji Banshoya, a Japanese planner who did work on Old Damascus and Old Aleppo, proposed a master plan for Aleppo in the late 1970s. The main idea for him was to end the abandonment of the Old City by connecting it to new quarters. Unlike his predecessor André Gutton, who wanted to cut off the Old City from west to east, Banshoya's main proposal was to reconnect both old and new, while at the same time preserving the historic quarters (see Figure 2.2).

Banshoya became the chief urban planner in Aleppo, producing a new master plan for the city in 1975.[20] His plan proposed smaller streets and parking lots behind the khans. According to historian Kosuke Matsubara, "Banshoya tried here to manage both conservation and activation of the old urban fabric at the same time."[21] The modernist influence of Michel Écochard, one of the prominent colonial urban planners, with whom he had worked, was clear. Although he turned down the intrusive nature of the roads proposed by Gutton and the level of destruction of the old fabric of the city, nonetheless he wanted to implement major changes. This led to the emergence of a grassroots movement in the city composed of urban notables and others who wanted to preserve the historic city and prevent the implementation of the master plan. Their movement garnered enough opposition to stop the implementation of the master plan and register the Old City as a historic monument at the UNESCO.[22]

1980: ALEPPO'S URBAN REBELLION

In the 1950s, the Muslim Brotherhood embraced a version of Islamic socialism without real economic depth. Their program was appealing to the middle and lower-middle urban classes, who were predominantly Sunnis. They had a small base in the rural regions, but no significant support.[23] After the Baath's rise to power in 1963, the Brotherhood retained its appeal among urban classes. Their primary constituencies remained the small bourgeoisie, shop owners, and artisans who had political and economic grievances against the Baath. The gradual eradication of the Sunni officers from the Baath further alienated Sunni urban classes, who believed rural and Alawite communities were becoming hegemonic in the political sphere. The combination of these factors led to the radicalization of a segment of the Muslim Brotherhood. The Hama branch, led by Marwan Hadid, organized a student protest in 1964, which turned into a city-wide revolt against the Baath regime. The revolt in Hama was mostly a reaction against the Baath program of nationalization of Sunni factories as well as the marginalization of Sunni landowners. By the mid-1970s, the divergence between Hadid's faction and the leadership of the Muslim Brotherhood grew wider. He consequently split from them and established his own group, The Fighting Vanguard of the Party of God. His followers believed "the impious Baath" had to be removed by force. In 1976, he was arrested and killed in prison.

The Damascus branch of the Muslim Brotherhood believed that working with the regime would benefit it. Its base was primarily composed of small bourgeoisie and merchants who aspired to build a close relationship with the regime that would benefit their economic interests. Palestinian historian Hanna Batatu explains, "[a] sort of de facto axis developed between military Alawis and the commercially minded Damascenes. The traders of Suq al-Hamidiyya never had it so good as in these years."[24]

The Hama and Aleppo branches of the Brotherhood were radically opposed to such an alliance. Several factors allowed their position to prevail, and pushed their party to adopt a militant approach toward the Assad regime. First, the Baath's military defeat against Israel in 1967 had a lasting negative impact on Syrian society. Second, the Baath support for rural classes at the expense of urban ones amplified the dissatisfaction of the latter. Third, large segments of the urban Sunni believed that the Baath policies benefited primarily Alawites and sidelined non-Alawites. When the Syrian government drafted a new constitution that cancelled the requirement for the president to be Muslim, the Brotherhood opposed it vehemently, demanding the reinstitution of the clause. Finally, Syria was facing a sharp economic crisis in the early 1970s, combined with an especially high inflation on basic commodities and rent.

After the Syrian army invasion of Lebanon in 1976, the Muslim Brotherhood's and leftist parties' opposition to the Assad regime intensified. The Muslim Brotherhood embarked on a campaign of assassinations, and in 1980, it organized a general strike, which led to the closing of the Market in Aleppo. The Fighting Vanguard killed more than 80 cadets, mostly Alawites, at the Military Artillery School in Aleppo, which triggered a wave of sectarian violence. The regime retaliated with a campaign of violence against the city. The Brotherhood was powerful and rooted in the city, and could challenge the regime in Aleppo in ways that it could not in Damascus. After the academy massacre, the Vanguard put pressure on the Muslim Brotherhood to become more involved in the confrontation against the regime. Historian Tine Gade notes, "The security situation deteriorated to the extent that people started to flee from the worst-affected cities, and large sectors of Aleppo, Syria's second largest city, were reported to be 'out-of-bounds' for the government's forces."[25]

In March 1980, Aleppo and Hama shop owners shut down the city for several days with a successful strike to protest the price controls imposed by the government to deter inflation. The Syrian regime deployed the 3rd Division and the Defense Brigades, and began a brutal crackdown on the city that resulted in several massacres. General Shafiq Fayad told his soldiers to "kill a thousand people a day to rid the city of the Muslim Brother vermin."[26] Residents and random passers-by in three districts, al-Masharqa, Bustan al-Qasr, and Suq al-Ahad, were rounded up and summarily executed (see Figure 2.3). The regime punished these quarters in particular because they did not show enough opposition to the Muslim Brotherhood. When the regime's military campaign finally ended, several thousand Aleppans were killed and injured, and thousands more were thrown in jail.

After the military campaign, the regime devised plans to alter the urban fabric of the city to prevent another rebelion in Aleppo. Analyzing Aleppo's urban plans, Jwanah Qudsi writes,

> Gutton's plan was met with resistance, and seldom executed when it was published. However, elements of it were adopted in the 1970s and 1980s by the new Baath regime, when the Old City's topographical and social fabric gave refuge to a resistance movement: the Muslim Brotherhood. The new arterial roads that were sliced through the Old City were essential for the deployment of armies through the area (not unlike Haussmanian roads in post-revolution Paris), and with time, allowed for the degradation and destruction of its patrimony.[27]

In the mid-1980s, due to the liberalization of the economy and the failure of land reform, many poor peasants began migrating in large numbers to the suburbs and informal settlements in Aleppo. Investment Law No. 10 of 1991 allowed Syrian and foreign businesses to invest in the industry and bring in hard currency without state control. These economic policies produced a new elite class that had strategic relationships with the regime. In addition, the Syrian government allowed middlemen to take over agricultural co-ops and state-owned lands and operate them, which exasperated small peasants and agricultural workers. The middlemen and large landowners gained from the new economic changes at the expense of the middle and poor classes.

Meanwhile, the pauperization of rural classes pushed many families to migrate to Aleppo. Many came from villages located to the north of Aleppo, such as Huraitan, Anadan, Tal Rifat, or Maraeh, and moved into poor neighborhoods such as Salaheddine, Seif al-Dawla, or Sakhour (see Figure 2.4); while those coming from Darat Izat lived in Kalasseh. These patterns of migration from Aleppo's countryside formed specific clusters within Aleppo that would have major implications when the revolt began in 2011. When rebels in Northern Aleppo began protesting and later fighting the Syrian forces, they were in constant contact with their families and friends in Aleppo. This explains to a certain extent how revolutionary knowledge was shared between the countryside and Aleppo, and how the geography of the revolt developed in the city.[28]

The second wave of migration was mostly due to the drought that lasted several years, from 2006 to 2010. This was combined with another wave of liberalization that Bashar al-Assad initiated in 2000. The economic implications of these policies on Aleppans were significant. On the one hand, the isolation of Aleppo due to its role in the 1980 rebellion was finally ended. A class of Aleppan bourgeois began emerging and taking advantage of the Syrian regime's openness toward their city. On the other hand, the economic liberalization and neglect of the agrarian sector produced a precarious rural class that moved in large numbers to the peripheries of the city to search for new opportunities.[29]

THE PROTESTS AND GRASSROOTS MOVEMENT IN ALEPPO

Aleppans organized their first protest on March 25, 2011, just a few days after the initial demonstrations in Damascus and Dara'a. Aleppo shopkeepers organized two successful general strikes in June.[30] June 30 became known as the "Volcano of Aleppo," and protests took to the streets in at least ten different locations. A few weeks later, on August 17, protesters reached Saadallah al-Jabiri Square, Aleppo's Tahrir Square, in large numbers for the first time. The largest protest to date, however, was during the burial of Aleppo's Mufti, Ibrahim al-Salqini, on September 6,

2011, when protesters marched in the Old City and chanted "Better death than humiliation!" At that point, protests, many of which were spontaneous, were organized on a daily basis at Aleppo University. Lawyers and the Bar Association issued a statement to denounce the violence of the regime, and held a protest at the Palace of Justice that was brutally repressed by security forces. Tansiqiat (local coordinating committees) were created in the most active districts of Aleppo to coordinate protests, as well as provide medical assistance to the injured. Hospitals had become too risky to go to, since the security forces had a permanent presence there and were arresting anyone they suspected of protesting. The Kurdish youth created their own Tansiqia, while university students established an Aleppo branch of the national Free Students Union. The residents of Sakhour and Marjeh districts, both in east Aleppo, were the first to create their own committees. These Tansiqiat organized large central protests as well as small, fleeting ones (known as "flying protests") to avoid the security forces' repression.[31] One of the focal points for the protesters was Saadallah al-Jabiri Square (see Figure 2.13), which is located in the city center, outside the Old City. The regime turned it into a military point while protesters were attempting to reach and occupy it.

As the protests grew in size and the Syrian regime's repression intensified, two opposing groups gradually formed. The first coalesced around the state's repressive institutions (security branches, Syrian army, various militias, and Shabiha) and the new Aleppan business class connected to Rami Makhouf's networks.[32] The Aleppan traditional merchant and industrial classes were split, because many were reluctant to oppose Assad due to the violence they had witnessed in the 1980s. In addition to these groups, Christian and Sunni religious leaders with whom the regime had cultivated a close relationship for decades mobilized their communities against the revolt. The protesters were mostly composed of the poor classes, who lived in informal settlements and the marginalized suburbs of east Aleppo, a segment of the shopkeepers and merchants, and university students.[33]

As the repression of protests intensified and massacres were committed in different regions in Syria, the peaceful uprising gradually turned into a military confrontation between the Free Syrian Army (FSA) and the Syrian army and the militias allied to it. In February 2012, insurgents began liberating villages in the north of Aleppo governorate. FSA fighters began appearing and defending protests in certain neighborhoods such as in Sukkari or Salaheddine (see Figure 2.4).[34] On July 19, 2012, the FSA entered Aleppo and took several neighborhoods; and in the following months, they liberated around 60 percent of the city, mostly districts located in the east.

WEAPONIZING DEMOGRAPHICS

To counter the protests and undermine the uprising in Aleppo, the Syrian regime weaponized demographics by splitting the population along different axes: religious, ethnic, tribal, and economic. These divisions were both political and spatial, and were instrumental in the defeat of the insurgents in 2016. One of the goals of the regime was the purification of west Aleppo, where protests and any forms of dissent were harshly repressed.

In the following section, I examine spatial violence in Aleppo through the concept of urbicide. Nurhan Abujidi explains that the literature on urbicide can be divided into three main areas. The first one includes authors such as Marshal Berman and Stephen Graham, who suggest that urbicide should be understood as violence against urban forms. They note that the city symbolizes pluralism and cosmopolitanism and its destruction is by definition an attack on these values. Berman explains how the demolition of urban forms after World War II led to the destruction of ways of life in the inner city. The construction of new urban infrastructure to create the suburbs undermined the inner city. For this school, Le Baron Haussmann is the ultimate example of the urban planner who did not mind destroying entire working-class neighborhoods to secure the pacification of what he considered an unruly class. In the same vein, other scholars such as Martin Shaw perceive urbicide as an anti-city war. The main historical example for him is the destruction of cities during the Balkan wars. This does not mean that rural areas would not experience any violence. For urbicidal processes to operate, the destruction of rural areas is necessary since this is where the roots of resistance are located.[35]

The second group points to the main effect of urbicide as the exclusion in the city. In this case, urbicide is defined as lack of plurality or heterogeneity that takes place. The destruction of the urban fabric is equivalent to undermining diversity in the city. Abujidi explains that "Urbicide for [Martin] Coward refers to the destruction of buildings in order to destroy shared places, spaces, and heterogeneity rather than limiting it to the destruction of urbanity conceived as specific ways of life in particular cities."³⁶ The main issue with that definition of urbicide is that in some cases the homogeneity of certain urban spaces is the main target as Derek Gregory demonstrates in the case of the Palestinian camp of Jenin, which was targeted by Israeli forces not necessarily because it represented a form of heterogeneity that needed to be fought at all cost. On the contrary, the main reason was that the camp needed to be destroyed primarily because it represented a way of life that is antagonistic to the Israeli way.³⁷

Finally, there is a third trend that intellectuals Stephen Graham and Derek Gregory best represent. These scholars explain that urbicide is used against those who inhabit threatening spaces and as such can be annihilated.³⁸ These authors explain that most wars take place in urban areas and as such, they are designed to target groups who operate within urban forms. According to this perspective, urbanism has been militarized to meet the challenges of modern warfare in post-colonial sites such as Iraq, Afghanistan, or Palestine.

Nuhran Abujidi proposes yet a different definition of urbicide. She explains that there is a direct and an indirect urbicide. The direct type targets the urban either to annihilate the enemy or to destroy body politics. There is also an indirect type that does not entail the destruction of urban forms but should be understood as the continuation of urbicide by other means. An example of that is the Israeli continual war against Palestinians that entails a segmentation of the territory and the positioning of the checkpoints to hamper the circulation of Palestinians and maintain a tight control over them.

Abujidi proposes to study urbicidal processes in three different stages. The first one revolves around policies concerning planning an urban space. These policies could be part of an ongoing process of destruction of certain spaces and populations. It requires the demonization of these groups before the urbicidal policies could be applied in the spaces they inhabit. The second stage is usually more rapid, visible, and destructive. It consists of actions, mostly urban warfare, that the state utilizes to destroy the targeted groups. The last stage, according to Abujidi, concerns the afterlife of destruction. It revolves around the effect of urbicide on the population and urban texture. What the author emphasizes by proposing such a definition is that urbicide is a process with its own temporalities and velocities rather than a one-time action.³⁹ Urbicidal processes are about ending diversity in the urban fabric. The regime was able to implement a clear separation between east and west. The spatial tactics utilized by the regime in Aleppo can help us understand the way the urban fabric can become an extension to a military strategy. The regime's weaponization of demographics prepared the stage for the deployment of urbicidal processes in Aleppo. The government's demonization of specific groups was a prelude to the destruction of the spaces they inhabit. In what follows, I examine the regime's spatial discourses about Aleppans and their neighborhoods, and how such rhetoric guided the geography of violence in the city.

Minorities

The regime produced a sectarian narrative about Syrian society that it instrumentalized against various social and political groups. As mentioned above, the war against the Muslim Brotherhood in the 1970s and 1980s allowed the regime to label any political opponent as Islamist terrorist. This narrative became dominant again during the current rebellion. Assad and his government made it clear that no one could oppose the regime, and that those who did were labeled Takfiri,[40] Wahhabi terrorist,[41] or foreign agent. To taint the image of the revolution and intensify the sectarian strive, the security branches released more than 1,000 Islamists from Sednaya prison in 2012,[42] in addition to thousands of petty criminals. The goal of the regime was to turn a popular rebellion into a Sunni war against minorities.[43] Four ex-prisoners in particular had major destructive impacts on the revolution. The first, Abu Khaled al-Suri, was a close aide to Osama bin Laden, and was behind the Madrid terrorist attacks of 2004. The second, Amr "Abu Atheer" al-Absi, was in the leadership of ISIS when he was appointed as governor of Aleppo by al-Baghdadi. The remaining two, Zahran Aloush and Hassan Aboud, formed and led the two largest jihadist military groups in Syria.[44] Most importantly, while the Syrian regime was releasing jihadists from Sednaya, it was also throwing thousands of secular and media activists in prison, many of whom died under torture.[45]

Aleppo is a microcosm of the Syrian social and political fabric, where different religious and ethnic groups have lived together for hundreds of years. The regime built al-Hamdanieh in the 1970s for Alawite officers and their families, many of whom worked in the adjacent military facilities (see Figure 2.5). In the 1970s and 1980s, the regime deployed a level of violence which Syrians had not witnessed before. The Muslim Brotherhood's sectarian violence against Alawites led to the migration of many of them to coastal cities, where they felt safe. In 2011, Assad wanted minorities to believe that the current revolt was a repeat of sectarian violence that took place in past decades. He used a politics of fear to mobilize the Alawi community and prevent the formation of alliances between Syrians of different religious backgrounds. As a result, many Alawi families living in Aleppo panicked in 2011 and left the city.[46]

Al-Shaar, a district located east of the citadel, was bombed on a regular basis for four years. The systematic destruction of the area was part of a plan to turn it into an Alawi or Shia district and gradually change the sectarian composition of east Aleppo, which is primarily Sunni. The regime is repossessing some of the houses and plans to give land to the Sunni owners elsewhere in Aleppo.[47]

In July 2012, when insurgents entered the city, many Sunni middle and upper classes opposed the move. They wanted to avoid the destruction of Aleppo, which was the fate of other cities that the FSA had liberated. The Syrian regime's message was clear: communities that allow insurgents to enter their districts would be bombed indiscriminately, and the entire population would be punished. In addition, the Crisis Group reported, "Aleppo's urban establishment views jihadis as a socio-political threat, an expression of the rural underclass' revolt. Secularists fear the spread and, ultimately, imposition of more conservative mores."[48] Since the middle and wealthy Sunni Aleppans lived in western Aleppo, it was easy for the regime to impose itself in these districts and isolate the rebellion in the eastern section. The regime solidified its control over these areas when jihadist groups, including al-Qaeda-affiliated Jabhat al-Nusra, chose Aleppo as their point of entry to the Syrian rebellion. In that regard, Aleppo's secular and non-jihadist groups were less dominant than in other cities. Al-Qaeda executed several suicide attacks in the city early on. The first Sharia courts emerged as early as August 2012.[49] They were created in reaction to the Unified Judicial Council (UJC), which these groups perceived as being too secular. Despite their opposition, all other groups in Aleppo signed a declaration supporting the UJC.[50]

Syrian security propagated a sectarian narrative in various ways, including the Syrian Electronic Army, which was active on social media.[51] It infiltrated protests to chant sectarian slogans, committed violence against regime forces and blamed protesters, and killed political opponents.[52] Activists explain that, in the early days, they would sometimes hear sectarian slogans such as, "Christians to Beirut, Alawis to the coffins," but would silence infiltrators who were chanting them with counter-slogans such as, "One, One, One, the Syrian people are one!" or "Syrians want freedom!"⁵³ Most Christians living in Aleppo tried to stay neutral, while a segment supported the regime either for pragmatic reasons or out of fear of sectarian violence. However, many young Christians supported the rebellion, especially Aleppo University students who witnessed the violence of security forces in their campus. In October 2012, American journalist Clare Morgana Gillis interviewed a young Christian man who participated in protests in Aleppo against Assad. She writes,

> Nearly all Aleppo's Christians live in just two districts and both are still controlled by the Assad regime. "There's a security checkpoint just under my house," he says, "and the road to get here is very dangerous." But attending anti-regime protests allows him to speak his mind freely, as few other Christians have been able to do during the last 18 months.⁵⁴

Before the uprising, 250,000 Christians were living in Aleppo, which was roughly 12 percent of the Syrian Christian population. The repeated attacks on Christians by jihadists and undercover regime security gradually made it very difficult for many to empathize with the rebellion. The regime was successful in instilling a climate of fear among Christian communities. Aleppo Christians live mostly in four neighborhoods, namely, Suleimaniya, Jdeideh, al-Midan, and Azizieh (see Figure 2.5); and when insurgents entered the city in July 2012, it was relatively easy to prevent them from taking these spaces. The regime maintained the climate of fear and distrust through sectarian conflict. For example, two Orthodox bishops from Aleppo were kidnapped in April 2013.⁵⁵ It is unclear whether the opposition or the regime was behind the kidnappings.⁵⁶ The nephew of one of the bishops explains, "I have to say that kidnapping of the two bishops serves only the Syrian regime."⁵⁷

The Armenian community living in Aleppo was generally opposed to the revolt. In 2012, Armenian religious leaders issued a statement in which they rejected the military confrontation and asked their community to embrace a neutral position. To punish them, the regime took away a parliamentary seat that had been allocated to Armenians since 1971.⁵⁸ Researcher Ara Sanjian explains, "[d]uring the first weeks of protests, most Armenians remained extremely loyal to the regime. Many of them willingly participated in officially sanctioned pro-Assad marches."⁵⁹ Several reasons explain the Armenian community's anti-revolt sentiment, which is more pronounced among them than other Christian groups living in Aleppo. First, Turkey's support for certain opposition groups made it difficult for Armenians to have a working relationship with the FSA. Second, the Syrian regime was successful in tarnishing the image of insurgents among minority groups, who believed most insurgents were radical fundamentalists. Third, Assad allowed Armenians to have their schools and clubs, and as such many felt they would lose their cultural autonomy in the event he was toppled.[60]

There are no accurate statistics about the number of Armenians living in Aleppo, but in the early 2000s, there were approximately 50,000.[61] Many were concentrated in two neighborhoods, al-Midan and Suleimaniya, both of which are in northwest Aleppo (see Figure 2.5). Christian youths formed a militia to defend churches by recruiting members from the Boy Scouts and the neighborhood. The militia was initially allied to the Kurdish Democratic Union Party (PYD)[62] (which shared its antipathy toward the Turkish-backed groups), but gradually sought funding and weapons from the regime. A small number were actively fighting alongside Assad forces, and created or joined pro-regime militias such as the Syrian Christian Resistance, the Martyrs of St. George Brigade, and the Warriors of Christ's Aleppo.[63] As the fighting intensified and members were killed, large segments of the Armenian community became more resentful and sectarian.[64] The Armenian genocide was instrumentalized to denigrate the rebellion. An Armenian website wrote "[t]he Syrian Armenians call Islamist militants 'Ottoman terrorists' and say that they are defending their lands and their compatriots from the same forces that committed Armenian genocide in 1915."[65] Thousands of Armenians chose to leave their neighborhoods and traveled to Armenia, when insurgents entered the city. The state propaganda was successful in dividing communities and weaponizing them against each other.

Informal settlements and poverty

The agrarian counter-reform under Hafez al-Assad and his son impoverished the peasantry and pushed many to migrate to the city. The creation of the "Arab Belt" at the border with Turkey displaced more than 100,000 Kurds, many of whom ended up in the suburbs of large cities such as Aleppo. In addition, the Syrian state went through two phases of liberalization: the first one was initiated in the mid-1980s and intensified immediately after the collapse of the Soviet Union, in 1991, while the second began when Bashar al-Assad seized power in 2000. The mismanagement of water resources for several decades and droughts in the late 2000s amplified the impact of economic reforms. The combination of these economic and political factors produced poverty on a large scale. In the case of Aleppo, the impoverished overwhelmingly stayed in the poor districts and informal settlements.

The overlap between the liberated areas of east Aleppo and the informal housing is striking. The insurgents controlled the vast majority of informal settlements (see Figure 2.7). The same thing can be said about the split between affluent and poor neighborhoods. None of the wealthy districts were under the control of the opposition (see Figure 2.6). This lack of social/class diversity gave the regime an important advantage. It was able to split the population along class lines and convinced the middle- and upper-class Aleppans that it would be in their interests to stand against what the regime portrayed as an invasion of rural Aleppo. In the end, the warning of the regime became a self-fulfilling prophecy. There was a takeover of Aleppo from the periphery. Many FSA fighters were poor and came from villages that surround Aleppo to the north and east. An FSA fighter explains, "[w]e liberated the rural parts of [Aleppo] province. We waited and waited for Aleppo [city] to rise, and it didn't. We couldn't rely on them to do it for themselves so we had to bring the revolution to them."[66]

A map of Aleppo's formal and informal housing provides a spatial understanding of how the regime devised a plan to control the city (Fig. 2.8). The density of the urban fabric in informal areas makes it difficult to control through a classical military force. As these areas are inhabited by the poor classes, it was easy for the regime to demonize and isolate them in the eyes of many middle- and upper-class Aleppans.

Youth and Aleppo University

In 2011, half the Syrian population was below the age of 25, 15 percent were unemployed, and large segments were poor. Evidently, the youth and university students were a source of worry for the regime, which utilized several strategies to prevent their participation in the revolts. The Mukhabrat[67] put pressure on families to deter their children from taking part in the protests. In the early days, the regime identified individuals with high symbolic capital and designated them as middlemen. They played a central role in the negotiation between families and the regime. To release an activist from prison, the middleman would be the guarantee to the regime that the young activists would cease all political involvement. The middlemen would then use their leverage (the power to release prisoners) to mobilize families and communities against protesters.

Aleppo University became a focal point for the security forces because it represented a real danger. It had a central location, and was the only space in western Aleppo where spontaneous and planned protests would take place on a regular basis (see Figure 2.4). The first protest was on April 13, 2011, one month after the initial protests in Dara'a. It lasted five minutes before Baath students hijacked it and turned it into a pro-regime rally.[68] The university was a unique space where students from different regions and diverse backgrounds discussed the uprising, organized protests, and refused to be silenced by the security forces. An Armenian student explained that the violence of security forces she witnessed at the university changed her mind about the revolt and the Syrian regime.[69]

The dormitory, which could accommodate up to 10,000 students, was the largest in Syria. Students from different regions, such as Raqqa, Homs, and Hasakah, as well as northern Aleppo, studied at Aleppo University and lived on campus. Many of them had lost relatives and friends during protests in their regions of origin, and had seen the FSA liberate their villages. As such, the university was a revolutionary space that pushed against the conservatism of the city. By November 2012, the students' legal team had identified 99 students who had been killed by security forces during protests. The Baath party office on campus was turned into a headquarters for the Shabiha and the security forces. Students from west Aleppo could not protest in their district because of the many checkpoints and community pressure in their neighborhoods. Instead, they would organize protests at the university or join the ones in south and east Aleppo, in working-class neighborhoods such as Salaheddine and Sakhour (see Figure 2.4).

On May 3, 2012, as the protests became larger and louder, security forces shot at the students and killed four.[70] A few weeks later, when a UN delegation visited Aleppo University as part of an investigatory tour, the students organized the largest protest in Aleppo, gathering around 10,000–15,000 people, and flew the independence flag on campus before being attacked by the Shabiha.[71] In the following days, security forces attacked students in the dormitory and wrecked their rooms, and finally closed the university before the end of the academic year. The last time Aleppo University had been closed by the regime was in 1980, when the city was under siege.[72]

The students kept protesting, despite the violence and constant threat, until the Assad forces bombed the university in January 2013 and killed 87 students. Journalist Malak Chabkoun writes,

> Students who witnessed the attack said they saw regime planes shelling the university, but what is significant here is what Shabiha did immediately after the attack which killed at least 87—campus security closed the university's gate, trapping students inside, and Shabiha began an impromptu protest in support of Bashar al-Assad—a protest amid a scene of bodies and blood, further terrifying the students who had witnessed the attack and preventing them from getting medical attention.[73]

The attack ended the cycle of protests and organizing in the only space of dissent in west Aleppo, and reinforced the boundaries between east and west. Some of the students moved to east Aleppo, while many left the city. After the bombing of the university, the level of violence against east Aleppo only intensified.

Tribes of Aleppo

The French colonial power weaponized tribes against nationalist and progressive groups operating in Syrian cities. When the Baath rose to power, it marginalized and revoked their privileges due to their collaboration with the colonial power. Syrian tribes couldn't control or manage the land anymore, as the customary law (*urf*) had been abolished.[74] The Baath believed the tribes were hindering its political and economic programs. During the Islamist rebellion in the mid-1970s to early 1980s, Assad altered the Baath's relationship with the tribes. He understood that tribal alliances could benefit the regime when faced with an existential crisis.

Tribal solidarity is a very effective tool to deploy against political opposition. The Syrian regime has always had two discourses toward tribalism. On the one hand, the Baath party considered them a residue from a past that needed eradication. On the other, tribes were instrumental in maintaining Assad in power. To reward their loyalty, larger tribes were given permanent seats in parliament. Researcher Haian Dukhan writes,

Hafez al-Assad used his patronage network with the tribes and unleashed their power to check the Islamists. Despite its national slogans of "no sectarianism" and "no tribalism," the Syrian regime did not hesitate to seek the aid of the tribes to suppress the uprising in 1982 in Hama, the stronghold of the Muslim Brotherhood.[75]

The Busha'ban tribe, who are from the Raqqa governorate, was sent to Aleppo to crush the Brotherhood rebellion. Assad used them to police the borders between Syria and Iraq and prevented the Muslim Brotherhood from smuggling weapons from Iraq. After 1982, they were rewarded with additional seats in parliament. Bedouin tribes loyal to the Syrian regime were used again against the Druze revolt in 2000[76] and the Kurdish rebellion in 2004; they killed protesters in both cases.[77]

When the uprising began, a growing number of bureaucrats, soldiers, and security agents defected. To make up for the disintegration of the state, the regime relied on identities and praxes that pre-date the nation-state. In that context, tribal, religious, and regional identities played an increasingly important role. The regime could not simply trust soldiers who might defect at any moment because they were from a region or a tribe that rebelled against Assad. It used some of the reliable tribes that it had tested in the past and had rewarded multiple times for their loyalty. Not only did these loyal tribes help to crush protests, but they also prevented their own members from participating in the revolt. The unconditional loyalty of some leaders ended up splitting certain tribes. According to Edward Dark, the al-Berri clan is the regime's primary enforcer in the city. "They don't even try to hide it and openly boast of receiving weapons and arms from the regime," he says. "There are lots of others too, usually convicted criminals involved in various smuggling or drugs. They were offered pardons and funds in order to help the security forces in the crackdown."[78] There were around 30,000 Shabiha in Aleppo, many of them recruited from the tribes.[79] Al-Berri, the largest and most powerful tribe in Aleppo, was instrumental in crushing the revolts, arresting activists, and killing political opponents. In addition, many tribe members lived in Bab al-Nayrab and Salihin, both of which are located in east Aleppo, and as such could police these areas effectively (see Figure 2.8). When the FSA entered Aleppo in July 2012, there was an agreement of non-aggression between it and al-Berri, which was violated by the latter. The FSA retaliated by brutally executing al-Berri's leader, Zeno Berri, and expelling al-Berri members from east Aleppo.[80]

Through this confrontation, the FSA removed one of the regime's last outposts in east Aleppo and the boundaries between east and west were reasserted once again.

Kurds

Kurds are the largest minority in Aleppo and live primarily in two districts, namely, Sheikh Maqsoud and Ashrafieh (see Figure 2.5). Kurds migrated to Aleppo in the 1970s–1980s from Afrin and Kobani, two Kurdish cities in the north, near the Turkish borders. To prevent Kurds from establishing their own nation-state, the Baath tried to create an "Arab Belt" on the border with Turkey. As a result, many Kurds were displaced from their villages, lost their land, and were replaced by Arabs who were brought from other areas.[81] Historian David McDowall writes, "A major socio-economic consequence of dispossession in the Arab Belt was increased Kurdish labour migration mainly to Damascus and Aleppo in search of work."[82] Most Kurds preferred to stay in these districts, primarily to avoid Arab discrimination toward them. These densely populated urban spaces are located in the northeast of Aleppo and have large sections of informal settlements. Infrastructure hardly exists in Sheikh Maqsoud, where there is no high school and only two primary schools, while some streets are unpaved. Since there are very few local jobs, the residents, many of whom work in the service economy, have to cross government checkpoints on a daily basis.[83] Middle- and upper-class Kurds live in Sabeel and Syriaan, two neighborhoods that are majority Christian.

In September 2012, the PYD and its military arm, The Popular Protection Units (YPG), controlled the district and evicted other Kurdish groups, including Yekiti, from the city. The PYD is a branch of the Kurdistan Workers' Party (PKK), which is led by charismatic leader Abdullah Öcalan, who exercises great influence on Kurdish politics in Turkey and Syria. The PKK was originally a Marxist–Leninist party, but more recently embraced democratic confederalism because its founder was inspired by the work of political theorist Murray Bookchin, who theorized about autonomy, confederalism, and anarchism. The PYD opposed the Syrian regime as well as the uprising, and instead concentrated on implementing the principles of democratic confederalism in Northern Syria and the Kurdish districts of Aleppo.

The Syrian regime's shelling of Sheikh Maqsoud on September 6, 2012, killing 21 Kurdish civilians, severed the relationship between the two.[84] Opposition forces attempted to control Ashrafieh, but were pushed back by YPG forces without much fighting. The FSA preferred to negotiate a non-aggression agreement with the PYD to keep the focus on the regime. The district has a strategic location, since it is located on a hilltop in between insurgent and government areas. In addition, it is situated near Castello Road, which became a central node for the opposition and the only point of access to east Aleppo. Thus, it was vital for the FSA to have a good relationship with the PYD.

Tensions between the PYD and Syrian opposition, however, gradually grew, for several reasons. The Kurds were fighting ISIS in the north and progressively became the dominant power in the region. The Turkish government formed a coalition of Syrian opposition forces to fight and prevent the Kurdish forces from consolidating their gains. These confrontations in Northern Syria—between Arabs and Turkmen on the one hand, and Kurds and their allies on the other—had catastrophic consequences on the relationship between Arabs and Kurds in Aleppo. As Sebastian Gonano reports, "In response, various rebel groups repeatedly and indiscriminately shelled Sheikh Maqsoud, killing and wounding hundreds of civilians."[85] In addition, Islamist groups in Aleppo, including al-Qaeda, had an antagonistic relationship with Kurds due to their progressive and secular politics. Opposition forces imposed a siege on the Kurdish district, leading to the deterioration of the living condition and shortages in water and electricity.[86] Amnesty International issued a report in May 2016 that shows opposition forces committed war crimes against the Kurdish population in Aleppo.[87]

When the Syrian forces reached Castello Road in July 2016, the YPG helped them to impose a siege on east Aleppo. In November, the cooperation between the regime forces and the YPG became more evident.[88] The Baath regime has oppressed the Kurds for decades by displacing them from northern regions and forming the "Arab belt" over a distance of 280km to separate Syrian Kurds from Turkish ones. It stripped hundreds of thousands of Kurds of their Syrian citizenship, crushed their rebellion in 2004, killed 36, and injured hundreds.[89] There were thousands of Kurds in Syrian prisons for political reasons.[90]

In addition, the regime repressed the Kurdish grassroots movement and killed Mashaal Tammo, one of its leaders.[91] Despite these atrocities, the Syrian regime capitalized on the divergences between opposition groups and the Kurds to crush the politics of life in the city. The strategic location of Sheikh Maqsoud was a vital asset for the regime and its forces. The PYG helped the regime impose the siege on east Aleppo in June 2016, and six months later take the territories controlled by the insurgents.

Palestinians

According to UNRWA, there were approximately half a million Palestinians living in Syria prior to the 2011 revolt. Displaced Palestinians fled to Syria in two phases: the first was after 1948 partition of Palestine; and the second was a result of the June 1967 Arab–Israeli War. The First Palestinian camp in Aleppo, al-Nayrab, is situated eight miles southeast of Aleppo and was established in a French barracks (see Figure 2.5).[92]

When Palestinians arrived in 1949, there were no independent Palestinian organizations. Around 20,000 refugees lived in the camp before the revolt. Its isolation and the lack of connections to Palestinian organizations made the task of controlling the camp rather easy for Syrian security. The second camp, Handarat, where 5,500 Palestinians live, was built in 1962 in northeast Aleppo (see Figure 2.5). By the early 1960s, Palestinians in the diaspora had their own organizations and were active in Palestinian camps in Syria. Thus, the population's living conditions in Handarat were superior to al-Nayrab, and their political autonomy was better compared to other Palestinian camps.[93]

The Baath party had a pan-Arab program and pro-Palestine rhetoric, but Palestinians living in Syria and Lebanon were as oppressed as any other Syrians. The Syrian army invaded Lebanon in 1976 and backed right-wing militias who massacred Palestinians in Tel-al-Zaatar, leaving several thousands dead or wounded. To control Palestinian politics in the Arab World, Assad created and funded several Palestinian organizations. These groups were based in Damascus and had no real autonomy but were effective in creating a rift in Palestinian politics and weakening the Palestine Liberation Organization.[94] In the 1970s, the military intelligence created Far Falistin (The Palestine Branch), which was supposed to support the Palestinian struggle by gathering strategic information but it quickly became one of the symbols of their oppression in Syria. The Palestine Branch is notorious for the gruesome torture that Palestinians who oppose Assad were subjected to for decades.[95]

In 2011, Palestinians, like other Syrians, began protesting throughout Syria and built an important grassroots movement in al-Yarmouk, the largest camp in the suburbs of Damascus. To neutralize Palestinians and prevent them from participating in the popular revolts, the Syrian government reactivated its discourse about Palestinian rights, while at the same time tasking Palestinian organizations loyal to the regime to crush the grassroots movement. In Aleppo, Muhammad al-Said, who was loyal to the regime and had a good relationship with Air Force security before the revolution, played an important role in preventing Palestinians' participation in the revolt. Before the revolt, he helped Palestinians living in al-Nayrab get the necessary permits to build houses or travel, and used that as leverage to prevent the emergence of independent Palestinian organizing in the camp. In 2012, he helped create Liwa al-Quds (The Jerusalem Brigade), a Palestinian militia with 500 Palestinians mostly recruited from al-Nayrab, Handarat, and al-Ramal in Latakia. The militia was Palestinian in name only, since most of its 3,500 members were not Palestinians.[96]

To prevent Palestinians from taking part in the revolts, Syrian security staged the kidnapping of 17 al-Quds fighters who were traveling in a bus; 14 of them were found dead a few days later. The regime then used the disfigured bodies to blame insurgents for the massacre and turn Palestinians in both camps against the revolt.

Unlike the vast majority of Palestinians in Damascus, who supported the revolt, most Palestinians living in Aleppo were either opposed to the revolt or silent. In March 2015, after liberating Idlib, FSA found pictures of two al-Quds fighters who were among the group slaughtered inside a security branch. The pictures proved that the torture took place inside the security branch, but the harm was done, and it was too late to alter the opinion of Aleppo Palestinians.[97] This is one example among many that illustrates the ways Assad instrumentalized the Palestine question to counter opposition to the government. In the end, neutralizing most Palestinians and creating a militia inside the camps in the north and south was strategic and important for the regime to maintain control over east Aleppo. Urbicidal processes operated through the demographic mosaic of Aleppo. The spatial distribution of the population was the foundation on which Assad built its military strategies. Some areas were used as buffer zones or obstacles while others played a role in the siege of east Aleppo. Overall, the regime instrumentalized the demographics of Aleppo to generate a well-calibrated politics of death.

FIXED CHECKPOINTS AND MOBILE MILITIAS

A taxonomy of checkpoints

The checkpoint is one of the most important technologies of death that the Syrian forces use in urban spaces. Their function is to segment the city into smaller sectors that are easier to control. There are different types of checkpoints in Aleppo that vary in size, the types of weaponry used, and their dangerousness. The large checkpoints with heavy weaponry, such as tanks or armored personnel carriers, are usually deployed at strategic points such as the entrance of the city or nearby security branches or military facilities. While most checkpoints are fixed, some of them are mobile and move to locations unexpected by the enemy. In addition, the regime deployed many checkpoints within west Aleppo to segment the territory and control the circulation of the population. The largest checkpoints, however, are in the buffer zones, the spaces between the regime and opposition. Their function is to prevent the infiltration of the enemy, but also to terrorize civilians who try to cross them. The smaller ones reaffirm the sovereignty of the state within the areas where they are deployed, while the larger ones delineate the boundaries of Useful Syria. The latter have a military purpose, while the former's main function is security.

The checkpoint is usually positioned at the border to maintain state sovereignty. Post-colonial scholar Stephen Morton notes,

> [c]heckpoints provide a spatial representation of the order of sovereignty, and the security apparatus of the state. Conventionally understood as "a barrier or manned entrance, typically at a border, where travellers are subject to security checks" (Oxford English Dictionary [OED]), checkpoints demarcate the contours of political geography, and police the movement of the human populations that traverse these boundaries.[98]

When the checkpoint is moved inside, within a territory, it evidently indicates a crisis of sovereignty.

The checkpoints around western Aleppo perform several functions that not only involve military calculations, but also the imposition of tariffs on commodities and people crossing them. Some of these checkpoints are terrifying, since Aleppans crossing them undergo a thorough search, intimidation, and harassment. Cell phones and computers are searched, including Word documents, Internet browsing history, pictures, Skype chatting history, etc. In many cases, civilians were arrested and disappeared because their browsing history shows they visited the Al Jazeera website or other media outlets critical of the regime.

The Syrian army controls the most restrictive checkpoints, which "are stationed along the highways entering the city and the two ring roads surrounding it."[99] The checkpoints located within western Aleppo are manned by the various security branches and are usually less restrictive. Caerus found that while the Syrian regime controlled only 35 percent of the neighborhoods in Aleppo in 2013–2014, it controlled 70 percent of all checkpoints.[100] The most lethal checkpoint in Aleppo is without a doubt Karaj al-Hijaz, and has become one of the symbols of the regime's violence (see Figure 2.9). It is known by residents as the "death crossing," and when insurgents were controlling east Aleppo, it was the only point Aleppans used to cross to the other side.

> Moving from the east to the western side of the city once took only a short bus ride. Now it involves navigating a labyrinth of side roads and as many as 20 checkpoints; an endurance test that can last between 10 and 16 hours. Most people don't bother.[101]

Residents try to avoid crossing Karaj al-Hajez until it is absolutely necessary. In 2013–2014, west Aleppo was under siege for several months, which drove the price of basic commodities up. The price of food and fuel went up respectively by 400 percent and 1,200 percent, which forced many residents to buy them in east Aleppo.[102] Some east Aleppans cannot find work in opposition area and are forced to cross the checkpoint on a daily basis and risk their lives. Others cross it to visit relatives on the other side. Two main snipers, in addition to two dozen auxiliary ones, control it, and often shoot civilians trying to cross to the other side. Residents documented that in April 2013, on average, snipers killed four people every day; and by February 2014, 213 people had been killed by snipers as they attempted to cross over to the other side.[103] The area in between is a no man's land, where the bodies of sniped civilians decompose for days without the possibility of removing and burying them. People have to walk approximately 400 meters to cross to the other side, and there are 22 snipers along the way; but the ones that target people the most are the ones positioned on top of the Municipal Building and Airways Building.[104] To avoid their bullets, the residents attempt to understand the snipers' logic and patterns but they often fail, as it is highly unpredictable. Even when Aleppans are successful in crossing the checkpoint, they are harassed and humiliated. One man explained that when the Syrian security and Hezbollah members found that he was from Masaken Hannano, one of the rebellious districts in east Aleppo, they humiliated him and put gasoline on his 13-year-old son and threatened to burn him.[105]

Aleppo Shabiha

The Shabiha ("ghosts" in Arabic) refers to the informal networks of thugs with whom the Syrian regime has developed strategic relationships since the 1970s. They are the eyes and ears of the state in the neighborhoods where they are based. To reward their loyalty, the regime allows them to engage in drug trafficking and smuggling of various commodities. They belong to various ethnic (Arab and Kurds), religious (Sunni, Shia, Armenians), or tribal (Berri and Mardini among others) groups in different cities. In Aleppo, Abu Ali Quzuk was the head of one of the large Shabiha groups, also called Lijan Sha'abia (Popular Committees). When the revolt began in 2011, he organized militias using a house located in Mouhafaza, an affluent district in the city center. After an assassination attempt against him, he left the neighborhood and moved elsewhere to keep the opposition activities away from the affluent areas. According to Saber Darwīsh and Mohammad Abī Samrā, there are three types of Shabiha in Aleppo: 1) the al-Berri clan, who had been assigned the task of maintaining order in the city by Hafez al-Assad since the 1970s; 2) the petty criminals, many of whom were released in 2011 and were organized in militias; and 3) the police force, who often joined Shabiha groups to suppress protests. The Syrian regime did not bother to fund them adequately; instead, it asked loyal Aleppo merchants to provide the necessary funds for the salaries of the Shabiha. Some merchants refused to pay when the amounts and frequency of the payment increased exponentially, so the militias began looting and kidnapping the children of rich merchants for ransom.[106]

The Berri Clan is part of the al-Jiss tribe, and they began working with the regime in the mid-1970s. Hafez al-Assad used them in the 1980s to repress the Muslim Brotherhood in Aleppo. They were given a permanent seat in parliament to reward their loyalty to the regime.

Some of them became Shia, and opened Hussainiat in Bab al-Nayrab and al-Salihin, where they have approximately 5,000 members (see Figure 2.8).[107] During the revolt, they formed Liwa Mouhamad al-Baqir along with other clans. Many tribal members from Aleppo rejected the official politics of their tribes and joined the Free Syrian Army.[108]

Kurdish Shabiha were also active in Aleppo; they are primarily composed of the Mardini tribe. In the 1930s, many Kurds, Arabs, Sunnis, and Christians migrated from Mardin in Turkey because of hunger and found refuge in Aleppo. They formed a close relationship with the state, and some became smugglers and had contacts with the security services. Many lived north of the Old City in Jabiri and Nile, which were constructed in the 1940s. Their leader was killed in 2014.[109]

The regime used a combination of checkpoints and Shabiha to crush the protests and any opposition in the city. Most checkpoints were fixed, but they sliced the territory into small and manageable sections. Shabiha were mobile and could be deployed rapidly and moved instantly from one district to another. In addition, unlike the army or foreign forces, who were unfamiliar with the geography of the city, Shabiha were often local and knew the city and its streets intimately. They had a deep understanding of the social networks, relationships, and spatial practices of Aleppans. They used mental maps instead of high definition military maps to repress opposition to the Assad regime. They were effective in the beginning, when there was need for quick and accurate intervention at Aleppo University, al-Jabiri Square, or isolated neighborhoods where protesters gathered. Urbicidal processes in the early period (2011–2012) were fine-tuned to suppress grassroots resistance in specific neighborhoods. After the liberation of eastern Aleppo (starting in July–August 2012), the regime began a campaign of systematic destruction of opposition areas.

THE DESTRUCTION OF EAST ALEPPO

The fighting between the regime and opposition forces in 2012–2013 led to a spatial separation between east and west Aleppo. In early 2013, the Syrian army and its allies stabilized the frontline around the wealthy areas and non-Sunni neighborhoods after losing most poor Sunni districts in east Aleppo. The weaponization of demographics in west Aleppo and the strategic positioning of the regime's military bases and security branches (see Figure 2.10) prevented opposition forces from moving further west.

Each side cleansed its territory from undesirable elements. On the one hand, the FSA killed the leader of the pro-regime Berri militia and forced its members out in July 2012. In addition, it attacked police stations and expelled the police forces from east Aleppo.[110] On the other hand, the Syrian army bombed the University in west Aleppo and killed many students to put an end to the wave of protests. Assad made it clear that no opposition would be tolerated in west Aleppo. In addition, the regime forces turned the topography of Aleppo into a lethal tool by weaponizing its hills, green areas, and river. The topography was utilized to further consolidate the separation between east and west. Syrian forces used Hannano barracks, located on a hill nearby Sheikh Maqsoud, to bomb and terrorize the Kurdish residents. Its purpose was to deter the Kurdish neighborhoods from joining the opposition.[111] To further reinforce the separation between east and west, the government militias kidnapped hundreds of civilians at checkpoints after the liberation of east Aleppo. In January 2013, more than a hundred were tortured and executed in a park in a regime-controlled area and thrown in the Queiq River. Their bodies were found downstream in Bustan al-Qasr, a district controlled by the FSA.[112] These examples demonstrate the extent to which the Syrian government weaponized the topography of Aleppo to subdue the population and maintain control over west Aleppo.

Once the borders between east and west Aleppo were erected, the regime began a campaign of systematic destruction of infrastructure in east Aleppo. The army used barrel bombs as a main tool to terrorize the inhabitants (see Figure 2.11). Urbanist Deen Sharp notes,

> [t]he army's aerial assault has relied on barrel bombs, low-cost cylinders filled with explosives, fuel, and steel fragments that are manually deployed from helicopters (Lloyd 2013). The bombs have become synonymous with the government's indiscriminate destruction of the Syrian urban landscape.[113]

The Syrian, and later Russian jets systematically targeted hospitals, bakeries, and gas stations.

Evidently, the primary motive of the urbicidal violence is to impose a geography of terror in east Aleppo. The massive destruction of east Aleppo with barrel bombs served several purposes. First, it created a climate of fear that demoralized the population and ultimately broke the resistance in Aleppo (see Figure 2.15). The targeting of hospitals had devastating effects on the population. A report published by the Atlantic Council in 2017 provides a detailed assessment of human loss and infrastructural damage. The authors explain,

> [a]ccording to the Syrian American Medical Society (SAMS), 172 verified attacks on hospitals or medical facilities were recorded across Syria between June and December 2016. Of those, 73 verified attacks—42 percent of the total—were recorded in the besieged, opposition-held half of Aleppo.[114]

The destruction of medical facilities, which the Syrian forces used extensively, is an intrinsic and vital component of urbicide.[115]

The second purpose of the massive usage of barrel bombs (which are inadequate for a military campaign since they are highly inaccurate) is to teach other liberated regions a lesson: no post-Assad Syria is permissible and those who violate this fundamental principle will pay a high price. Third, by driving the residents out of the city, the regime made it easier for its army and militias to invade it. Fourth, the government created an additional burden on the liberated territories by pushing large number of Syrians into them. The liberated areas were already facing an economic crisis and shortages of basic commodities. They had limited resources to meet the needs of their inhabitants and thus struggled when they had to share them with an increasingly larger number of displaced Syrians. Finally, the regime destroyed large sections of east Aleppo and uprooted the inhabitants in order to rebuild these areas according to neoliberal urban plans. The politics of post-conflict reconstruction is a continuation of urbicide by other means.[116] Countless Aleppans were unable to return to their homes because pro-regime militia members occupied them.[117] In many cases, the deeds of the estate were purposely destroyed to prevent the inhabitants from returning to their neighborhoods.[118]

VERTICAL POWER

Snipers played a crucial role in the Syrian conflict by initially preventing gatherings in public spaces, and later by terrorizing people living in areas controlled by insurgents. When protests erupted in March 2011, the Syrian regime positioned snipers on rooftops to target peaceful protesters and disrupt demonstrations. Government media claimed infiltrators, not Syrian soldiers, were shooting at protesters. Syrian army defectors, however, claimed the real perpetrators were the regime's soldiers. A sniper deployed to Izraa, near Dara'a on April 25, 2011 told Human Rights Watch,

> General Nasr Tawfiq gave us the following orders: "Don't shoot at the armed civilians. They are with us. Shoot at the people whom they shoot at." We were all shocked after hearing his words, as we had imagined that the people were killed by foreign armed groups, not by the security forces.[119]

Since the beginning, the role of snipers was primarily to terrorize civilians, and only secondarily to target military objectives. Historian Mirjana Ristic explains,

> [U]nlike Foucault's (1995) conception of the panopticon, the snipers' gaze in Sarajevo was not a disciplinary technology or normalization regime. Rather, snipers used the asymmetric visibility as a mechanism for the production of terror. A significant part of Sarajevo's public space was reconstituted into a "landscape of fear"—a network of dangerous and forbidden zones in which any mundane, everyday activity became potentially lethal.[120]

Likewise, sniping in Aleppo is about imposing a topography of terror and preventing a normal life in opposition-controlled areas. Countless witnesses have told terrifying stories about snipers targeting children playing in the street, civilians shopping, and even mentally disabled persons wandering aimlessly in the city after the bombing of the psychiatric hospital.[121]

Ristic identifies two types of snipers operating in Sarajevo, Yugoslavia during their civil war. The first was the leisure sniper, who was not from the capital, did not know its residents, and would shoot at any moving target—including children and other non-combatants. These leisure snipers were "'paid killers'—for whom the killing of Sarajevo's residents was a part-time job."[122] The second type was the considerate sniper who lived in the city before the war, and would avoid targeting residents. Instead, he used his bullets to shoot at empty buildings or broken cars. In some cases, he would alert residents when he saw their children playing in the sniper's alley.[123]

The Syrian regime was highly aware of the risks of deploying snipers, and more generally soldiers, in their city of origin. Since the massacres of 1980–1982, it recognized the need for bringing soldiers in from elsewhere to crush a revolt. Additionally, the regime often used one ethnic group or religious community against another. The regime's deployment of non-Syrian militias (Iraqi, Afghani, and Iranian) or armies (Iranian and Russian) to repress its own population addresses that specific problem. Bashar al-Assad crushed a Druze rebellion in November 2000 and a Kurdish uprising in March 2004 with the help of Arab tribes. It recruited tribes from northeastern Syria and soldiers from the coast and Damascus to massacre civilians in Aleppo and Hama in 1980 and 1982.

Most of the regime's snipers in Aleppo are of the leisure type. There is no precise information about the number of snipers deployed in the city, but they are probably in the hundreds, if not thousands. David Nott, a British surgeon who worked in field hospitals in opposition-controlled areas, explains that every day, snipers would target a different part of the body for leisure. He explains,

> From the first patients that came in the morning, you could almost tell what you'd see for the rest of the day, […] It was a game. We heard the snipers were winning packets of cigarettes for hitting the correct number of targets.[124]

Snipers were positioned along 10 miles of fighting front which went through intertwined spaces controlled by regime and opposition forces.[125] Activists created Facebook pages to provide information about the location of snipers and the alleys or neighborhoods they could target. One sniper, if well positioned, could control a large section of a neighborhood and keep people inside their houses for extended periods. Aleppans began putting up signs to alert residents of the presence of a sniper in a specific area.[126] Local journalists wrote countless articles about how to avoid a sniper, and what to do in the event someone was injured.[127]

Aleppo residents tried to understand the logic of snipers and develop strategies accordingly to avoid their bullets, but it proved to be a futile exercise. For example, a sniper could target a woman or a child without killing them. Then, when a resident tries to help, both the rescuer and rescued would be killed. In some cases, two snipers in different locations coordinate to kill a civilian. The first shoots at a target from one side to force her to escape, while the second kills her from the opposite side. To protect themselves, Aleppans used buses stacked on top of each other to barricade a street and block a sniper's line of sight.[128] Families sealed their windows with sandbags to avoid snipers' bullets, while others were forced to live in one section of the house as the rest was too exposed. The FSA positioned mannequins in specific locations to confuse snipers, and in some cases to determine their locations and inform residents. By examining the bullet's points of entry and exit in and out of the mannequin, it is possible to determine the angle from which it was shot and therefore identify the sniper's position.[129]

Snipers positioned on the tallest buildings in Aleppo were the most terrifying, since they could reach a much wider area and were difficult to eliminate (see Figure 2.12). For example, Aleppans suffered a great deal from the snipers positioned on the rooftop of the Municipal Building, which is the tallest building in Aleppo, with 23 floors. Syrian forces turned the city's entire bureaucratic structures into military outposts to target civilians and insurgents in east Aleppo.

As explained above, the death crossing in Bustan al-Qasr is one of the most feared points in the entire city (see Figure 2.9), because it is the only location Aleppans can use to go to the regime's side. Aleppo's inhabitants called it the "Corridor of Death," since many have lost their lives trying to cross it.[130] Doctors working in east Aleppo often tried to save the lives of those shot at the crossing. They estimate that 10,000 people used the crossing every day, and 15–20 civilians were killed.[131] Around 30 percent of the causalities in Aleppo were from snipers' bullets, while 50 percent were killed by barrel bombs, and another 20 percent died in combat.[132]

The study of urbicide in Aleppo requires an examination that goes beyond a two-dimensional analysis of a map. To understand the way snipers built a topography of terror in the city, I examine their networks and the spaces they covered through their well calculated positioning. Analyzing the significance of verticality, geographer Stephen Graham explains,

> Until recently, addressing such a question has been hampered by the dominance of remarkably flat perspectives about human societies in key academic debates about cities and urban life. In geography, especially, territory, sovereignty and human experience have long been flattened by a paradoxical reliance on flat maps—and, more recently,

aerial and satellite images – projected or imaged from the disembodied bird's or God's eye view from high above.[133]

A three-dimensional analysis is essential to comprehend the logic behind the government's deployment of vertical power in Aleppo. One of the important advantages the Syrian regime has over opposition forces is its total supremacy over the sky. This is combined with its control of the affluent districts and city center, where the high-rise buildings are (see Figure 2.12). Table 2.1 shows a list of these buildings with their heights.

Table 2.1 A sample of high-rise buildings in Aleppo

1	Ar-Rahman Mosque	246 ft
2	Municipal Building	≈271 ft
3	Shahba Sham Hotel	≈235 ft
4	Mirage Hotel	≈177 ft
5	The Citadel of Aleppo	≈164 ft
6	Syrian Airways Building	≈153 ft
7	Great Mosque of Aleppo	≈147 ft
8	Syrian Railways Building	≈141 ft
9	Aleppo Sheraton Hotel	≈118 ft

These high-rise buildings date from different periods, and represent different forms of power. Four types of buildings can be identified: 1) official/governmental buildings; 2) luxury hotels; 3) mosques; and 4) historical monuments. The first type symbolizes state power and its bureaucracy. The most prominent example of that power in Aleppo is the Municipal Building, which is located in the city center nearby Saadallah al-Jabiri Square, the square protesters attempted to reach countless times in 2011–2012 (see Figure 2.13). It is also a strategic location for snipers, since it overlooks opposition-controlled areas on three sides. The second type represents the power of the capitalist class, whether mercantile or neoliberal. The proximity of this elite class to the Syrian regime allowed it to invest its wealth in tourism and build luxury hotels. High-rises such as Mirage Hotel (previously Amir Palace Hotel) (see Figure 2.13) or Shahba Sham Hotel became strategic locations for snipers, and were frequently targeted by opposition fighters. They illustrate the organic relationship between Syrian bureaucracy and the Aleppan capitalist class.

The third type of vertical power is the religious institution. The tallest mosques, namely, Ar-Rahman and Hafez al-Assad, which were built in the 1990s and 2000s respectively, dwarf the Umayyad mosque, which dates from the eighth century (its minaret was built in 1090). The tallest mosques are all in the regime-controlled areas, and many of them were used for sniping. The Syrian regime built these mosques in the wealthy neighborhoods to polish its image and please the Aleppan pious community. Finally, the citadel represents several layers of historical sedimentation that the Assads inherited from the past (see Figure 2.14). It played a crucial role in the twelfth and thirteen centuries when Muslims used it to stop the Crusades' advances. When the FSA reached Aleppo, Syrian forces positioned more than 100 snipers to maintain their control over it and the surrounding areas. The opposition forces used several tactics, including digging tunnels around it to take it back, but were not successful. A fighter positioned in the area explains,

"There have been many attempts by the rebels to liberate the citadel," [...] But the high ground and high walls, and the bullets fired so easily through its arrow slits, [...] "make the mission to liberate it too difficult for now."[134]

CONCLUSION

The regime's control of western Aleppo in 2016 was not coincidental. It was the result of many factors, including the urban history of the city, the way demographics were weaponized, and the regime's overwhelming superiority in the sky. The fact that all the high-rise buildings were located in western Aleppo was not accidental either. These structures are the urban manifestations of political, economic, and religious powers, and as such they were located in affluent Aleppo. The combination of these factors gave the regime a major advantage over the insurgents and civilians living in east Aleppo. Against the Syrian forces' supremacy, residents in eastern Aleppo had limited tools to defend themselves.

Belgian urban planner Le Corbusier played an instrumental role in the colonial project in North Africa. He created urban forms in Tunisia and elsewhere to help the French colonial power maintain control over its colonies. Stephen Graham writes, "[Le Corbusier] celebrated both the modernism of the aircraft-machine and its vertical destructive power. 'What a gift to be able to sow death with bombs upon sleeping towns,' he wrote in his 1935 book *Aircraft*."[135] Le Corbusier wanted to destroy the Old City and its dense urban fabric because it is an easy target for bombers in time of war. He argued for replacing it with towers located far apart, and as a result, difficult to target by air.[136]

Le Corbusier's colonial mindset sheds light on the way urban forms, which seem benign in time of peace, can be weaponized in time of war. During the second half of the nineteenth century, wealthy notables and Christian families moved to modern Ottoman districts in west Aleppo. They left behind the walled city, its dense urban fabric, and unsanitary housing. Their new districts witnessed the erection of different types of high-rises, while the city they had left behind expanded into informal housing. The legacy of these urban processes, some of which are more than a century old, is still with us today. More importantly, these processes were weaponized by a criminal regime to suppress a popular revolt and kill civilians.

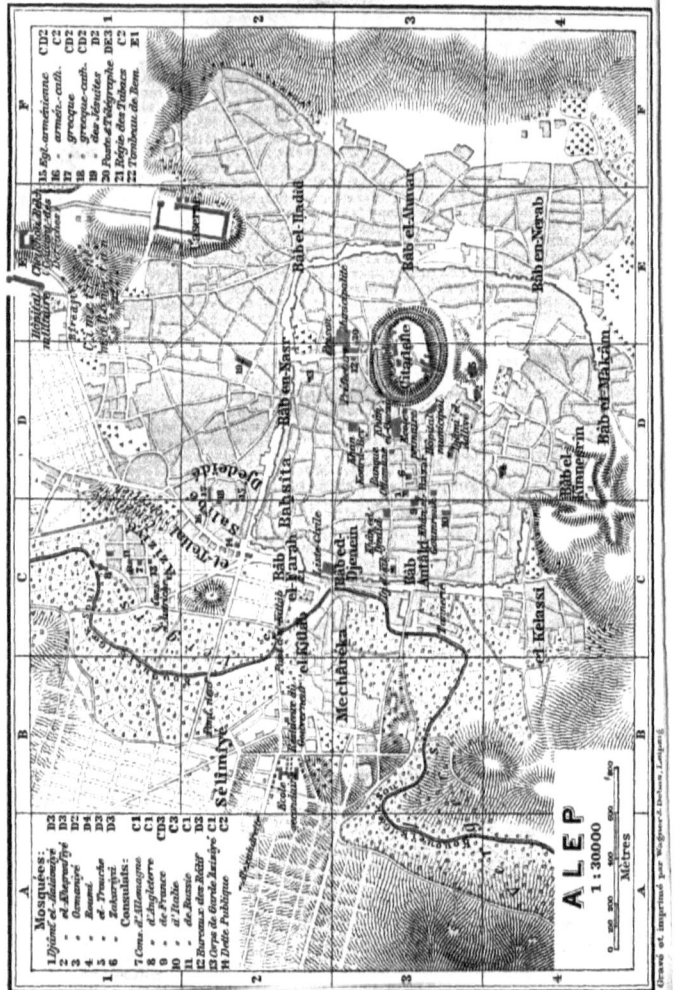

Figure 2.1 Aleppo in 1912, showing the newly-established and wealthy neighborhoods west and northwest outside the Old City. (Wikimedia. Accessed August 20, 2019)

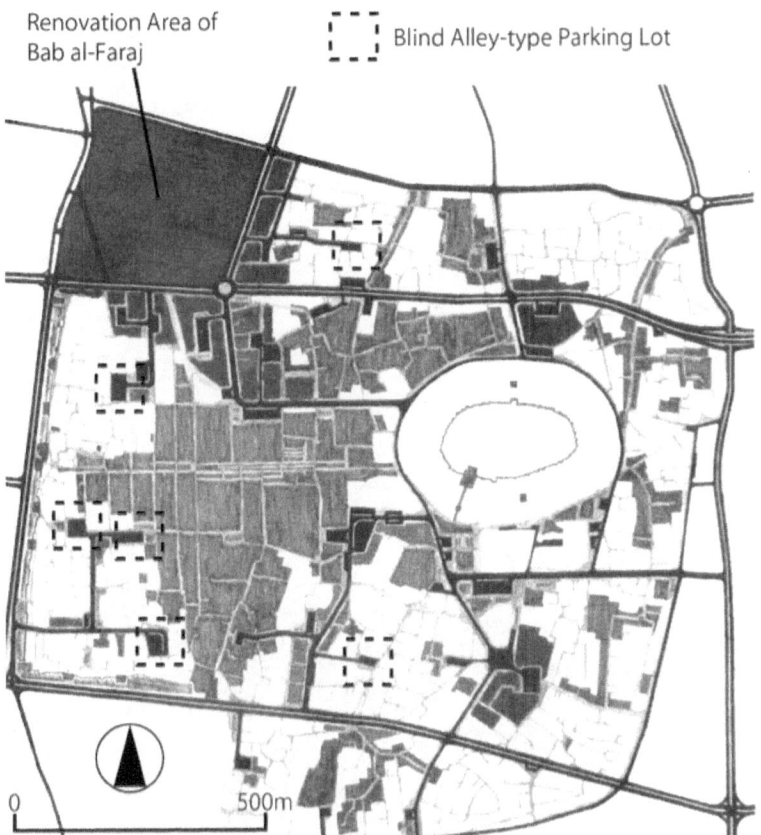

Figure 2.2 Gyoji Banshoya's proposed plan for Aleppo, 1973. (Banshoya, Gyoji, and Jean-Claude David. "Projet d'aménagement de la vieille ville d'Alep [Project of improvement for the old city of Aleppo]," *L'Architecture d'Aujourd'hui* 169 [1973]: 84–85)

82 · THE SYRIAN REVOLUTION

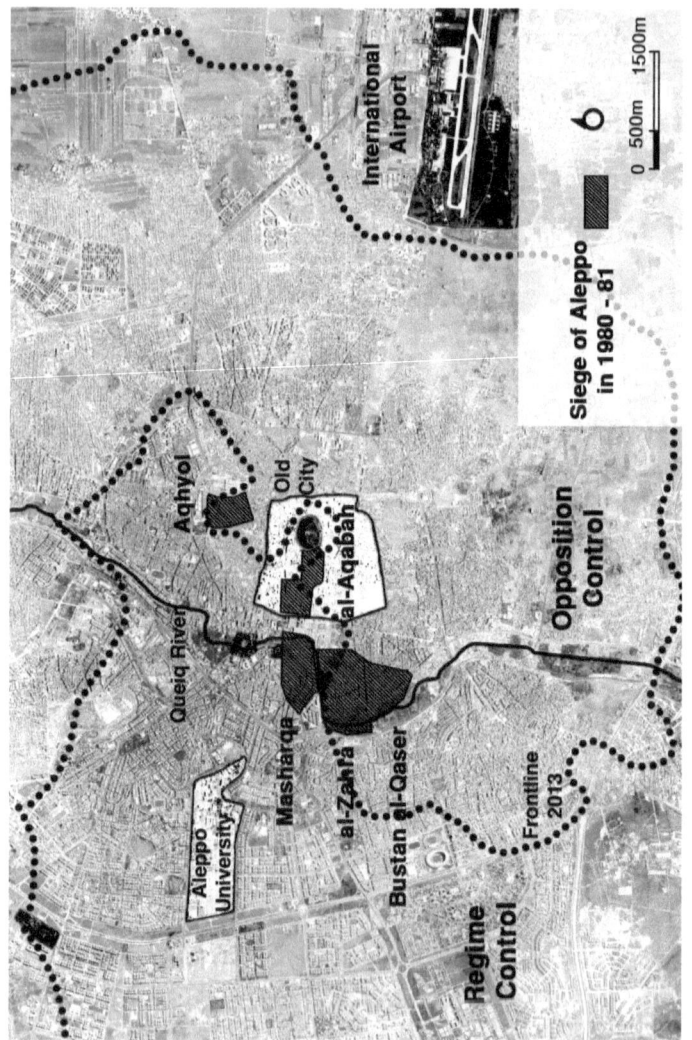

Figure 2.3 Aleppo's rebellion and neighborhoods' siege in 1980–1981. (Based on "Siege of Aleppo (1980)." Wikipedia. Accessed May 18, 2019. https://en.wikipedia.org/wiki/Siege_of_Aleppo_(1980))

THE GEOGRAPHY OF DEATH IN ALEPPO · 83

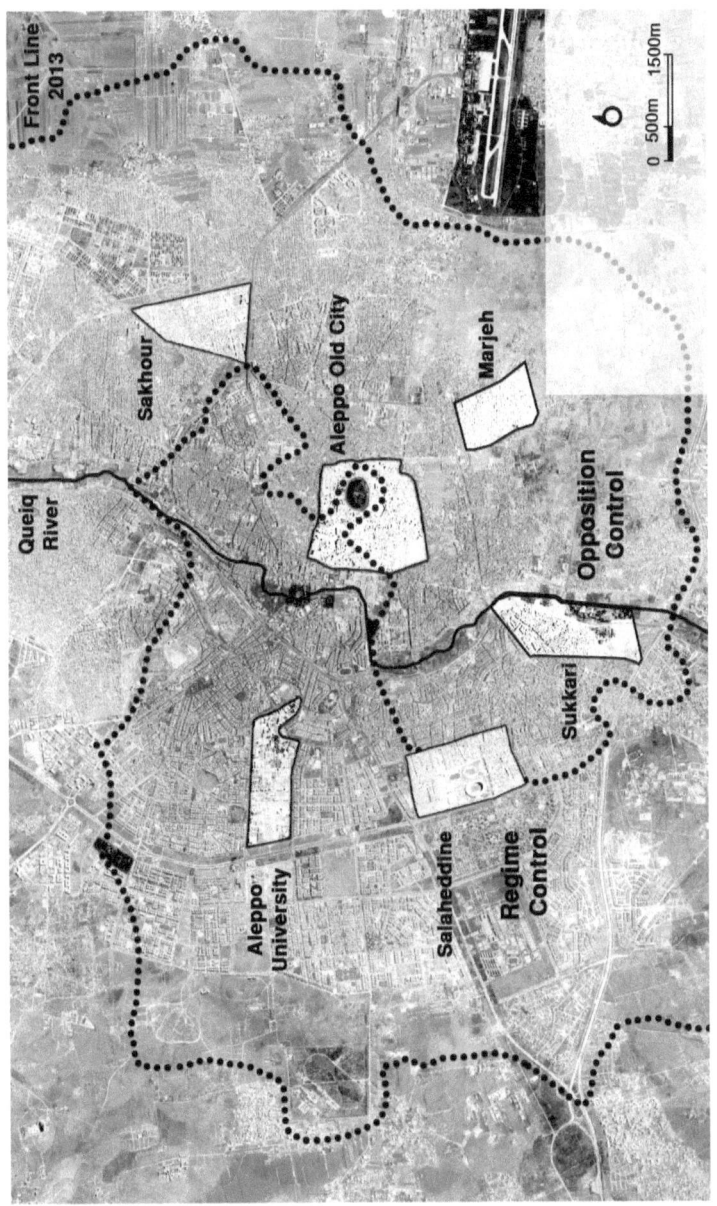

Figure 2.4 Protests in 2011–2012 and Frontline in 2013. Neighborhoods where most protests were taking place.

84 · THE SYRIAN REVOLUTION

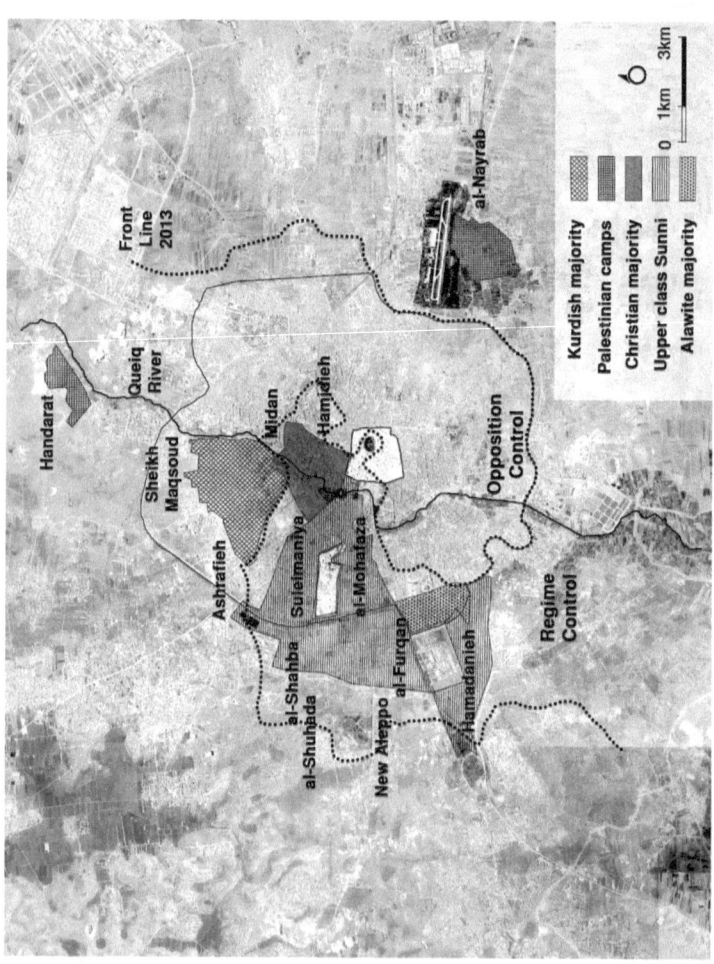

Figure 2.5 Aleppo's religious and ethnic composition.

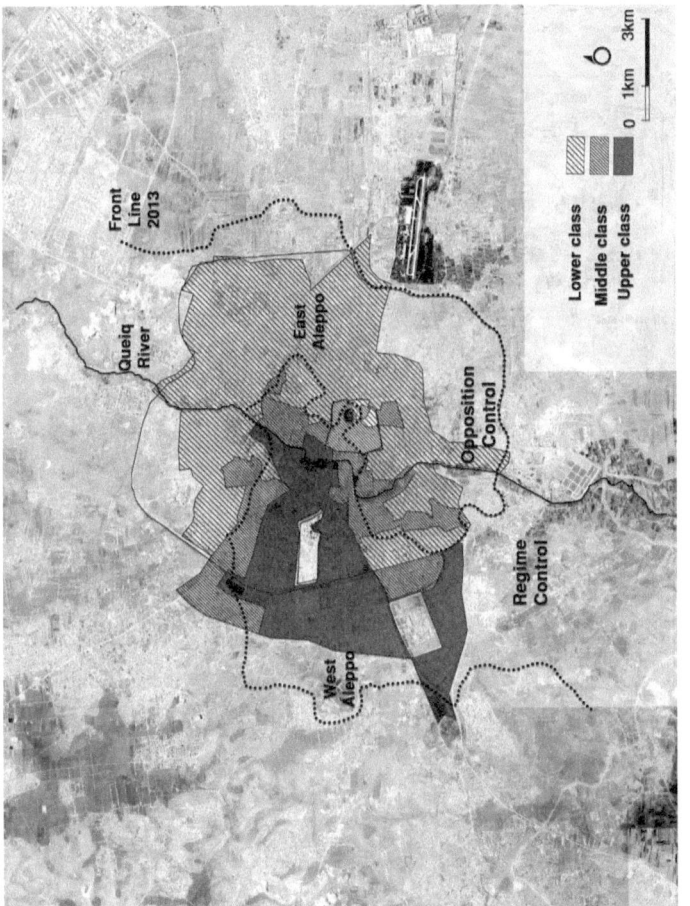

Figure 2.6 Socio-economic distribution in Aleppo. (Hussein Almohamad, Anna Lisa Knaack, and Badriah Mohammed Habib. "Assessing Spatial Equity and Accessibility of Public Green Spaces in Aleppo City, Syria," *Forests* 2018, 9(11), 706. https://doi.org/10.3390/f9110706)

86 · THE SYRIAN REVOLUTION

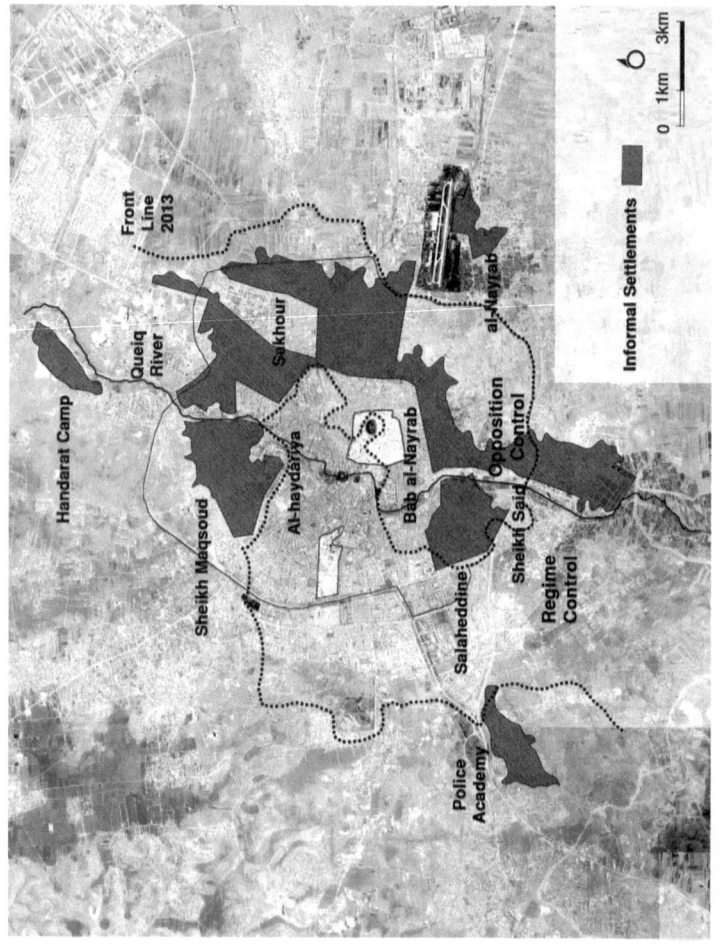

Figure 2.7 Informal settlements in Aleppo.

THE GEOGRAPHY OF DEATH IN ALEPPO · 87

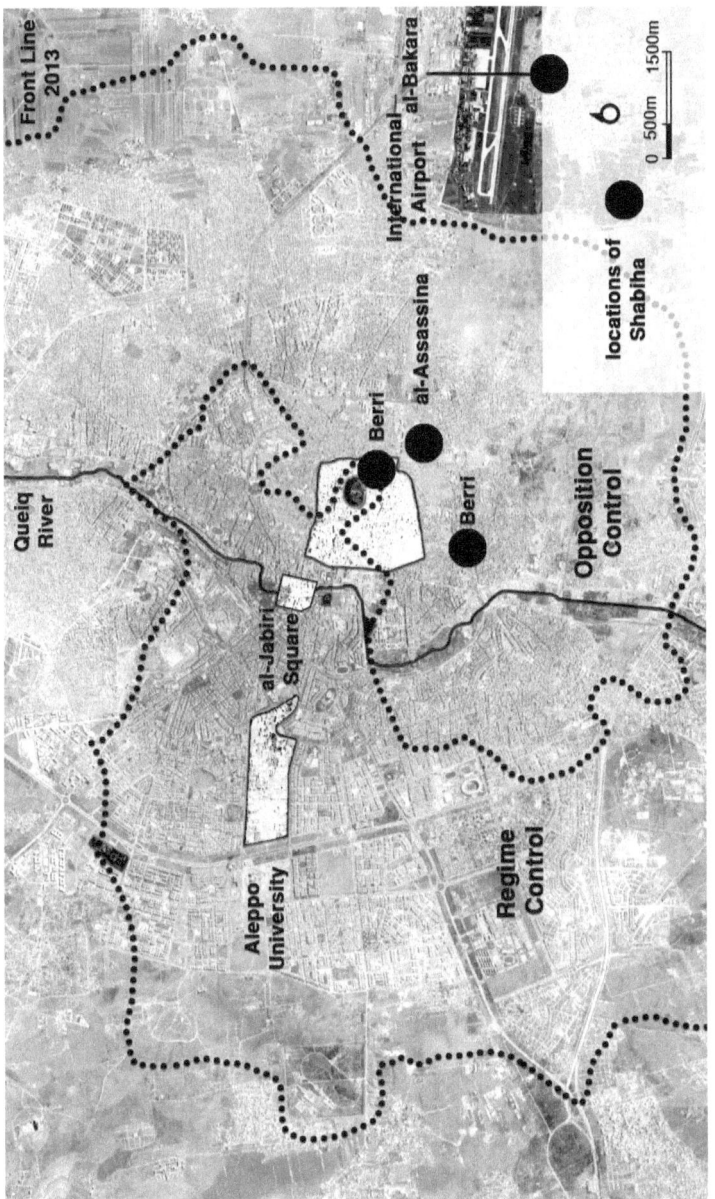

Figure 2.8 Neighborhoods where most Shabiha groups were based in east Aleppo before the revolt.

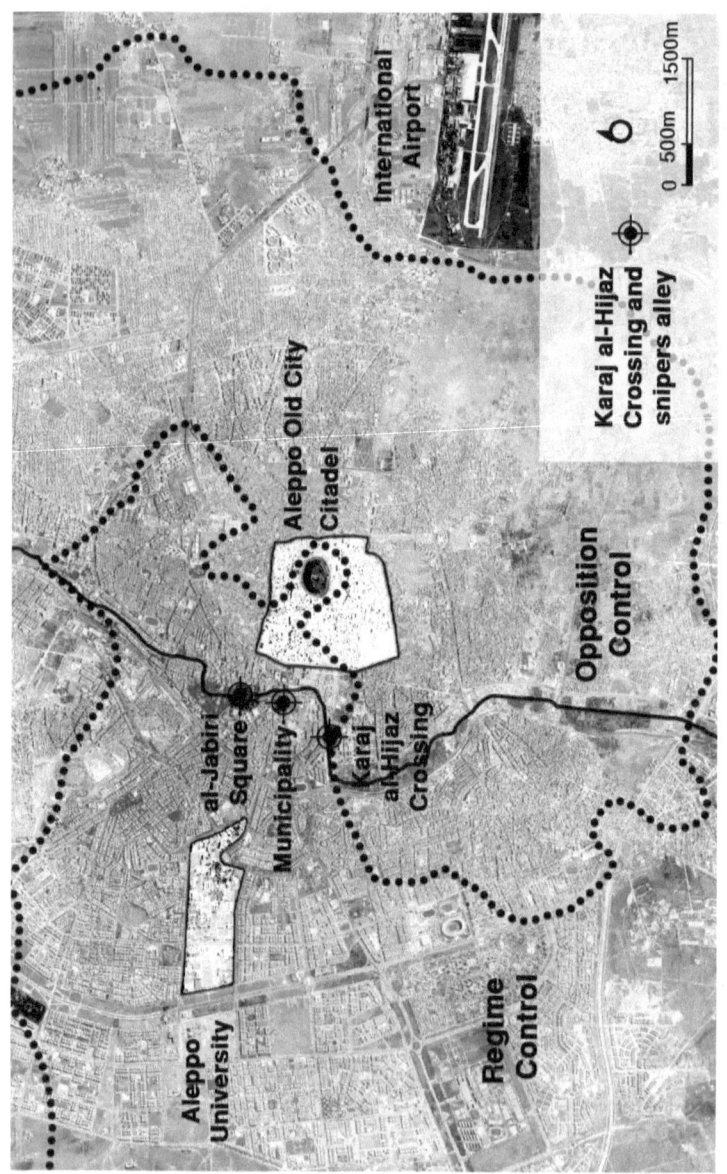

Figure 2.9 Snipers positioned on the Municipal building's rooftop targeted civilians crossing Karaj al-Hijiz checkpoint.

THE GEOGRAPHY OF DEATH IN ALEPPO · 89

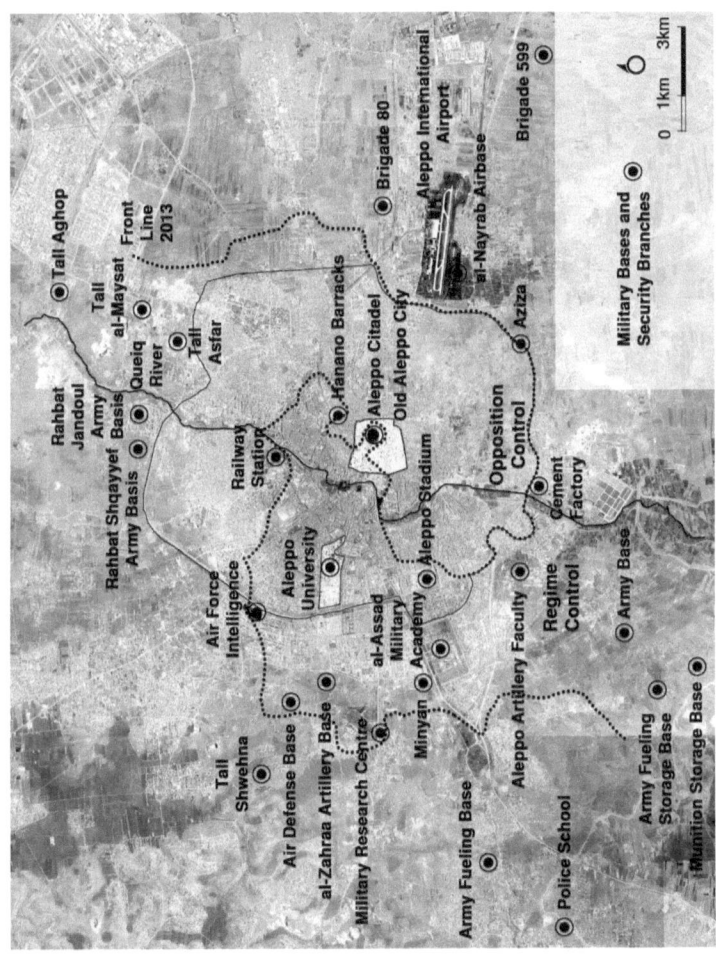

Figure 2.10 Military bases and security branches in and around Aleppo.

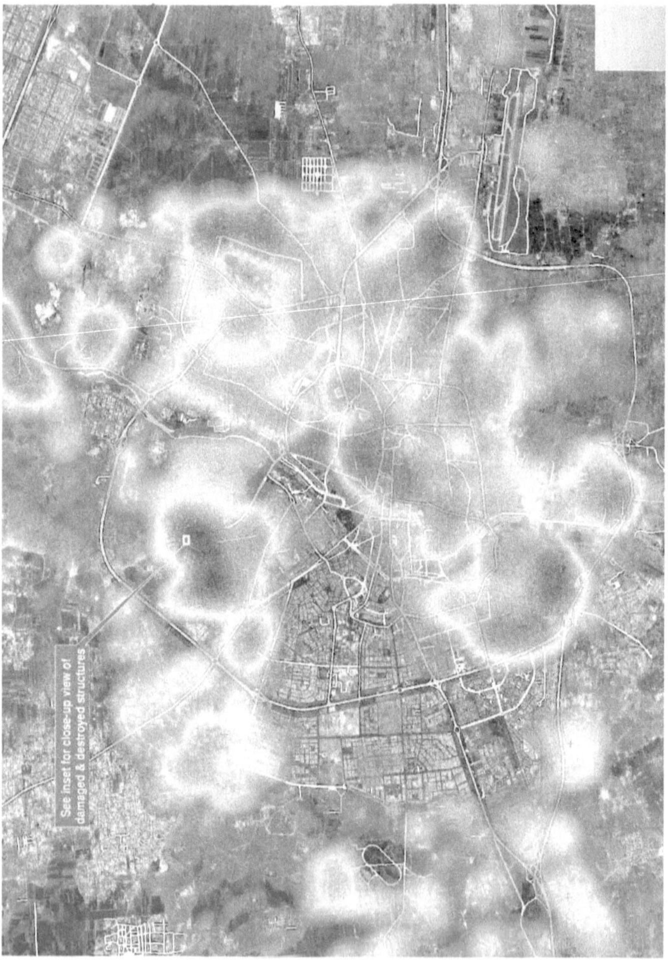

Figure 2.11 Damage density in the city of Aleppo (UNITAR map, December 2016). The map shows a clear contrast between regime and opposition-controlled areas. (UNITAR. Damage density in the city of Aleppo, Syria. *Unitar*. December 16, 2016. Accessed June 2, 2019. https://unitar.org/unosat/node/44/2508?utm_source=unosat-unitar&utm_medium=rss&utm_campaign=maps)

THE GEOGRAPHY OF DEATH IN ALEPPO · 91

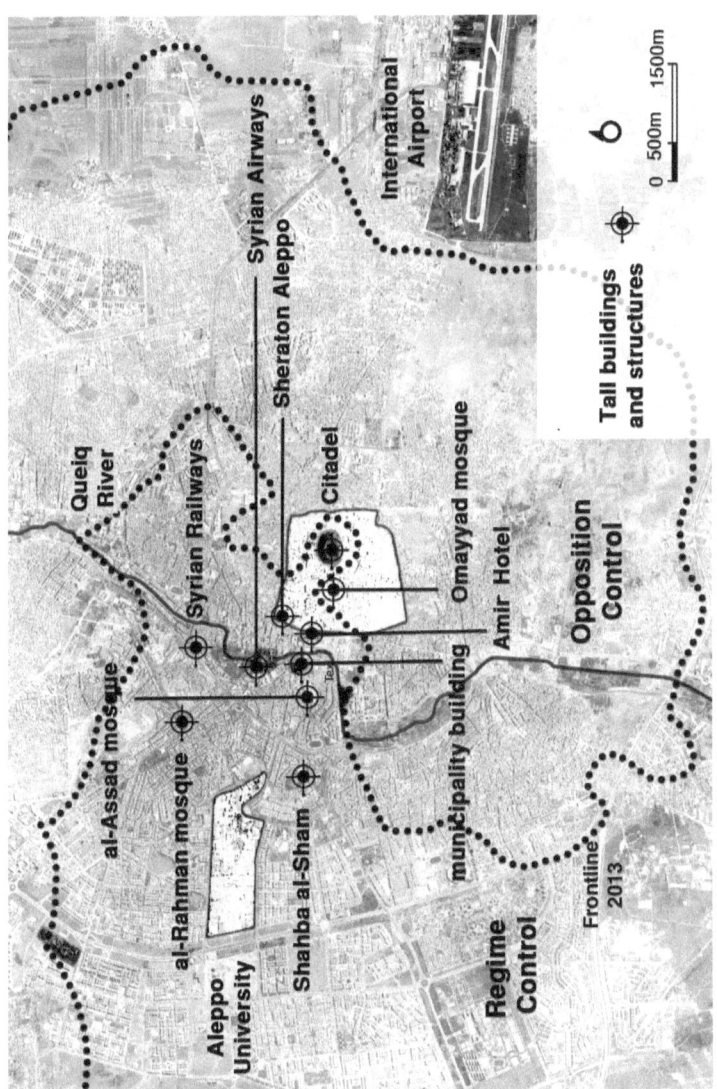

Figure 2.12 Locations of main snipers in west Aleppo.

Figure 2.13 Saadallah al-Jabiri Square that protesters attempted to occupy multiple times in 2011–2012. Pro-regime snipers were positioned on the rooftop of the Municipal building (on the right) and Amir hotel (where the picture was taken). Photographer: Roderick Aichinger.

THE GEOGRAPHY OF DEATH IN ALEPPO · 93

Figure 2.14 The Citadel overlooking Aleppo city. (Stock Photo. Photographer: Sean Sprague)

Figure 2.15 The politics of life: children turn a bomb crater into a swimming pool in Sheikh Saeed in Aleppo – August 2016. (AP via Aleppo Media Center)

3
Nation Against State: Popular Nationalism and the Syrian Uprising

> *[The Bourgeoisie has] come to power in the name of a narrow nationalism [...]; they will prove themselves incapable of triumphantly putting into practice a programme with even a minimum humanist content [...].*
>
> Frantz Fanon, *The Wretched of the Earth* (1963)

THE CURRENT SYRIAN REVOLTS

"One, one, one, the Syrian people are one!" In 2011, this was one of the most popular chants during protests. Syrians used it to counter the sectarian discourse of the regime. Arab nationalisms throughout the region have been reimagined and transformed by rebellious populations since the eruption of the revolts. Various actors in the region and beyond are trying to redefine the contours of official and popular nationalisms. Despots and protesters are using nationalism to mobilize the population around their respective political projects.

In addition, nationalism is critical in a region where a self-appointed Islamic State briefly dismantled century-old national borders and imposed what appeared to be a pre- or post-national state, depending on how one looks at it.[1] The Kurdish forces in Northern Syria and Iraq are also attempting to create a new reality on the ground based on nationalist sentiments.[2] The Syrian regime has suggested on several occasions, because it was unable to control the entire Syrian territory, that it is willing to fragment the country and create what it called a "Useful Syria," the most vital part of the nation, according to Assad's public statements.[3]

This chapter explores the significance of nationalism in the context of the Syrian revolt. It demonstrates that since 2011, Syrians have been developing a new form of popular nationalism.[4] They have done so by using nationalist narratives through revolutionary councils and other institutions, as well as via slogans, graffiti, media outlets, and other discursive and non-discursive tools. This chapter examines these emergent discourses and practices in Manbij, a city in the Aleppo Province of Northern Syria that has played a relatively important role in the liberated areas. This makes Manbij an excellent site for the study of popular nationalism.

Furthermore, the chapter shows that the establishment of a new national community is a vital necessity for resistance against despotism. This is why the Syrian regime is using every tool at its disposal to counter and undermine popular nationalism. The regime's violent efforts to do so are particularly visible in the liberated areas, where it fears the emergence of a post-Assad Syria that could threaten the Baathist state. The regime has used a combination of hegemonic and coercive tactics in its efforts, producing multiple counter-narratives, including the imposition of an official nationalist discourse to challenge popular nationalism. Since 2011, both nationalisms, popular and official, have competed for hegemony. These battles of narratives are taking place at the same time as the ongoing military confrontations on the ground.[5]

The decolonial thinking of Frantz Fanon, the Martinican intellectual and revolutionary theorist, can help us understand the anatomy of nationalism in the context of the Syrian revolt. In *The Wretched of the Earth*, Fanon predicted that Algerian independence would be incomplete after the military defeat and departure of the French army. He understood that Algerian nationalists who fought for independence would gradually become the middlemen between the native population, especially the peasant classes, and Western political and economic elites. His predictions about Algeria, and more generally about the Arab world, were sadly accurate, as the neocolonial political maps of the current era can attest. Fanon writes,

> The national middle class taking up the old traditions of colonialism, makes a show of its military and police forces [...]. We have seen that inside the nationalist parties, the will to break with colonialism is linked with another quite different will: that of coming to a friendly agreement with it.[6]

The Fanonian framework is useful in the context of this chapter. Syrian independence, like its Arab counterparts, was unable to reach its natural conclusion; it was aborted by a pan-Arab party in the 1960s, and since 1970 has been undermined by an autocratic regime. Greek scholar Anna

Agathangelou explains that the 2011 Arab revolts should be understood as a continuation of nationalist movements for national liberation that were aborted half a century ago.[7] Viewed from this perspective, the current revolts are against both post-colonial despotism and neoliberalism. Iranian-American intellectual Hamid Dabashi notes, "Pan-Arabism is not just a reversal of, and a reaction to European colonialism, it is also a replica and reproduction."[8] In this sense, Fanon's insight provides a strategic vantage point to understanding the ideological dynamics of the Syrian uprising.

This chapter's first section explores two vital moments in the history of Syrian nationalism. The first moment started in the early twentieth century, when Syria witnessed the emergence of popular nationalism. This early form of nationalism was instrumental in the struggle against French colonialism. The second crucial moment began in the early 1960s, when the Baath party seized power. The ruling party produced an official nationalism based on the writings of Michel Aflaq, Zaki al-Arsuzi, and Salah al-Din al-Bitar, which helped it maintain a position of dominance in public discourse. The second section examines the emergence of a popular nationalism since 2011 with a focus on Manbij.

FANON'S NEW HUMANISM

In *The Wretched of the Earth*, Fanon proposes decolonial tools to explore the question of nationalism. He identifies two types of nationalism. On the one hand, a bourgeois nationalism is utilized by African elites to build a hegemonic bloc against colonial powers. However, in the post-colonial period, this elite class became economically and politically subordinate to Western powers. The indigenous Arab elite strove to emulate the Western model, while simultaneously deploying pan-Arab and anti-imperialist discourses.

The second type of nationalism analyzed by Fanon in *The Wretched of the Earth* is anti-colonial. His analysis is based primarily on the experience of the Algerian Front for National Liberation and its struggle for independence. The Algerian struggle sometimes produced a Manichean worldview in which the colonizer and the colonized confront each other. There are no gray areas and no spaces in between.[9] British scholar Nigel Gibson explains that anti-colonial nationalism reached a deadlock in the post-colonial period. He writes, "Rather than developing new relations with the peasants and workers and genuinely involving them in the decision making process, nationalism regards them merely as the means to accumulate the capital needed for 'modernization.'"[10]

Fanon suggests that the way out of this binary framework is to envisage a third type of nationalism. Such a nationalism, which Fanon calls "new humanism," would consist of an authentic process of decolonization in the Global South. This form of nationalism would open the field of possibilities and produce different subjectivities. It is a nationalism that does not follow a script or a rigid ideological repertoire. Rather, new humanism is a nationalism of praxis that takes shape through people's struggles and everyday resistances. It furthers the political goals of the downtrodden. In addition, it is a nationalism that proposes a philosophy of mutual recognition and reciprocity.[11]

Prior to undertaking an examination of decolonial nationalism or the new humanism that is emerging in certain areas in Syria, an exploration of two important moments in the history of nationalism in the Levant is necessary. A concise historiography of Syrian nationalism is beyond the scope of this chapter.[12] Instead, the chapter proposes a short historical exploration to contextualize the current struggles against dictatorship. Paradoxically, a historical exploration of nationalism to explain emergent discourses seems almost antithetical to Fanon's philosophy. The Caribbean intellectual warns in the first page of *The Wretched of the Earth*:

> [D]ecolonization is quite simply the substitution of one "species" of mankind by another. The substitution is unconditional, absolute, total, and seamless. We could go on to portray the rise of a new nation, the establishment of a new state, its diplomatic relations and its economic and political orientation. But instead we have decided to describe the kind of *tabula rasa* which from the outset defines any decolonization.[13]

To justify his "*tabula rasa*" claim, Fanon notes that the violence of colonization obliterates pre-colonial cultures and prevents their reconstitution. For Fanon, nothing from the past is worth saving or preserving. One might suggest that he conflates dominance and hegemony, which are two distinct dimensions of colonialism. British scholar Neil Lazarus explains that domination of a colonial power over a people does not necessarily lead to the hegemonic imposition of colonial values and norms over subaltern groups.[14] In other words, colonialism does not always erase Indigenous culture, as Fanon might suggest.

While Lazarus' point is a valid one, it is possible to read Fanon as a historian of the present. In that regard, it is true that Fanon rejects the possibility of unearthing the past, but he opens a space for a reading of the past from the standpoint of a decolonial present. This is how one ought to read Fanon's call for a "history of decolonization." Lazarus writes:

> The immobility to which the native is condemned can only be called in question if the native decides to put an end to the history of colonization—history of pillage—and to bring into existence the history of the nation—the history of decolonization.[15]

It is this history of decolonization that this chapter attempts to present here. It argues that only such a history can unsettle and destabilize bourgeois and anti-colonial nationalisms that are both competing for hegemony in the current Syrian context. Unlike them, decolonial nationalism, or what Fanon calls "new humanism," and what is referred to in this chapter as popular nationalism, occupies marginal spaces in Syria. Instead of proposing an accurate reading of the multifaceted and heterogeneous emergent culture of the Syrian uprising, this chapter proposes a perspectival reading of "the history of decolonization."

EMERGENCE OF SYRIAN NATIONALISM

Before exploring the anatomy of popular nationalism in Syria since 2011, it is vital to examine two important moments in its recent history. The first focuses on popular nationalism in the early twentieth century, while the second analyzes state nationalism that emerged during the post-independence period. Unlike the state-sponsored nationalism of the Baath party, popular nationalism is grassroots, praxis-oriented, bottom-up, and liberatory. It first emerged to oppose the oppression of the ruling classes as well as colonial domination. The emergent popular nationalism since 2011 is polyvalent and can be observed in the sites of struggle, whether they are liberated or under the regime's control.

The new popular nationalism in Syria should be understood as both a rupture with the state-centric nationalism of the Baath party, and a re-imagination of the early nationalism of 1919–1920 and the Great Syrian Revolt of 1925–1927. Popular nationalism is operating a dual move that vacillates between delinking from an oppressive and centralized nationalism of the Baath party, which represents state terror, while simultaneously connecting with the democratic, decentralized, and multifaceted popular nationalism of the early twentieth century.

The emergence of nationalism in the Levant is a much-debated topic. Lebanese-American historian Philip Khoury has written extensively about it. He suggests that nationalism emerged in the region in reaction to the revolution of 1908 of the Young Turks. The revolution was a counter-move against the absolutist rule of Sultan Abdul Hamid II. It was followed by a politics of Turkification and centralization. The revolution of the Young Turks opposed any form of autonomy that the Syrian elite was requesting at the time. In addition, the notables of Aleppo, Damascus, Hama, and Homs felt that Istanbul failed to defend Muslim interests against European economic, political, and cultural hegemony. They reacted by developing an oppositional identity that Khoury qualifies as an emergent Arabism, but which was in reality an unstable and undetermined ideology, a combination of Arab, Muslim, regional, and local identities.[16]

Roger Owens refutes Khoury's analysis about the emergence of nationalism during the first decade of the twentieth century. He argues that Arabism in the early twentieth century was a combination of Islamism, regionalism, and loyalty to a town or tribe. According to Owens, Sharif Hussein adopted Arabism to defend his own interests, but was not really pursuing a project of Arab unification. His revolt was motivated by selfish political goals and an interest in maintaining political power.[17]

The revisionist historiography of scholars such as James Gelvin and Michael Provence shows that the nationalism of notables during the first decade of the century was elitist and did not affect popular classes.[18] These authors demonstrate that the notables could not have had any authentic nationalist aspiration, since they were demanding autonomy for Syria rather than complete independence from Ottoman rule. The question for Khoury and others is, why would notables develop and disseminate a nationalist discourse if they were not interested in independence? For these revisionist authors, nationalism emerged at least a decade later, as a reaction to the defeat of the Ottoman Empire at the hands of the allies during World War I. Gelvin shows that the penetration of capitalism in the region undermined traditional relations of power. Instead of the vertical ties that linked individuals to a family, a tribe, or a landowner, the new economy favored horizontal connections, and as such made national belonging an intuitive choice. The existence of these objective factors does not necessarily produce nationalism. Gelvin acknowledges the existence of a nascent Arab nationalist identity among Syrians during the war, but he also explains that this identity did not become a dominant paradigm until after the war. Popular nationalism, which emerged after the war, was countering the pragmatic and top-down nationalism of King Faysal.[19]

What is particularly important in the context of this chapter is that during these years, there was an unprecedented political mobilization among Syrians. Gelvin notes that in fall 1919, a broad collation of intellectuals, notables, lower-middle class religious dignitaries, *qabadayyat*,[20] and merchants was formed. They opposed Faysal's government, which they viewed as illegitimate and controlled by foreigners. They created democratic popular committees in many neighborhoods. Damascus had 48 neighborhood committees, each of which had representatives in the Higher National Committee. They also formed committees of national defense, not only to maintain security in their neighborhoods but also to protect Syria in the event of foreign aggression. For the first time, Syrians from all social backgrounds became directly involved in politics through popular committees. Syrians took matters into their own hands by creating a structure of democratic committees at the neighborhood, municipal, regional, and national level. These committees challenged the power of the central government and rendered local politics obsolete, while making national democratic politics a commonsensical choice. They performed a number of tasks, including guaranteeing a fair price for grain, recruiting volunteers for the committees of national defense, and providing relief to the displaced and the poor.[21]

The parallels between the popular nationalism of 1919 and that of 2011 are striking. Although explicit connections might not exist between the local coordination committees, which were formed in many neighborhoods starting in August 2011, and the popular committees of 1919, their appearance during periods of deep state crisis is telling. The popular nationalisms of the early twentieth and early twenty-first centuries both produced new institutions, cultures, and discourses at the national level. In both cases, nationalism was developing without concern about capturing state power. The main objective for both was to create a parallel structure to the state that could ultimately undermine the existence of colonial and neocolonial states. In both cases, popular classes in different regions were transcending localism and establishing connections with other groups at the national level. Fanon explains in *The Wretched of the*

Earth that nationalism is instrumental in developing an emancipatory discourse against occupiers. However, he urges his readers to transcend the seductive, yet destructive consequences of anti-colonial nationalism, and its aspiration to a Manichean politics. He writes,

> Only the massive commitment by men and women to judicious and productive tasks gives form and substance to this consciousness. It is then that flags and government buildings cease to be the symbols of the nation. The nation deserts the false glitter of the capital and takes refuge in the interior where it receives life and energy. The living expression of the nation is the collective consciousness in motion of the entire people. It is the enlightened and coherent praxis of the men and women. The collective forging of a destiny implies undertaking responsibility on a truly historical scale.[22]

The new humanism that emerges in peripheral territories is not simply oppositional to the colonial or dictatorial power. It also empowers people, and provides them with tools to tackle their everyday problems.

This first wave of nationalism in 1919 and 1920 prepared the terrain for the Great Syrian Revolt of Jabal Druze against French rule. The revolts, which took the French by surprise, lasted two years. The French were afraid that opposition to their rule would come from urban centers. Instead, the most steadfast resistance came from the countryside. The resistance of Shaykh Salih al-Ali in the Alawite region was not put down until late 1921. Resistance also came from the countryside of Aleppo, where Ibrahim Hannano led a very effective military formation. However, the staunchest and most sustained struggle was that organized by Druze communities in Hawran.[23]

The Great Syrian Revolt was a catalyst for the propagation of a Syrian Arab identity in the countryside. To counter the effective resistance in the countryside, French authorities produced a great deal of propaganda. They insisted that the revolt was not sustainable, because feudal elites could not convince Syrian peasants to join their cause. They also distributed leaflets to tarnish the image of the revolt by presenting it as sectarian and, as such, antithetical to non-Druze communities. Despite its intensity, the French propaganda was infective. For the most part, peasants were willing to set aside their class differences and join the battle against French colonialism. Druze peasants in the South and Sunni and Christian grain merchants in the Maydan Quarter in Damascus built an organic relationship through commerce. When French bureaucrats introduced new policies that threatened their livelihood, Syrians quickly organized a network of resistance that spanned from the Druze Mountain in the South to Damascus. Druze, Christians, and Sunnis were more than willing to cross sectarian lines and regional differences and join the resistance to protect their economic interests and livelihood.

It is important to note that the rebels used fluid discursive strategies to undermine French power. In some cases, they advocated for Muslim solidarity or tribal culture. While they highlighted class conflict in certain situations, they mostly used a nationalist discourse to oppose French rule, maintaining a poorly defined meaning of what it meant to be Syrian or Arab. This openness allowed different groups to join the rebels' battle and consider the combat against colonial power to be a main goal. The Great Syrian Revolt generated a blend of Syrian and Arab nationalisms based on loose meanings of patriotism, anti-colonialism, religion, and tribal honor.[24]

Michael Provence explains that the French lost no time in portraying the rebels as extremists, criminals, terrorists, and sectarian individuals who were concerned only with preserving the feudal system in Hawran. They used counter-insurgency and mass killing to quell the revolt. The aerial bombing was unprecedented at the time, and punished the entire population in the regions where the rebels took refuge. Al-Harika in Damascus was bombed for two consecutive days to drive the rebels out.[25] Military might was combined with a campaign to delegitimize the struggles of the revolutionaries. The French produced a counter-narrative that focused on the supposed sectarianism of Druze, the goal of which was to deter other religious communities from joining the struggle for independence. Provence writes,

> Sectarian conflict was a theoretical necessity for French colonialism in Syria, since the entire colonial mission was based on the idea of protecting one sectarian community, the Maronite Christians, from the predations of others. Without sectarian conflict, colonial justification evaporates.[26]

In addition, the French formed sectarian militias made up of Armenians and Circassians to ignite sectarian violence; but the rebels responded with a nationalist discourse that reminded Syrians that all sects, including Christians, Druze, Alawites, Shia, and Sunni, were "sons of the Syrian Arab nation." French authorities resorted to all forms of collective punishment and terrorization, including aerial bombardment, house demolition, public hanging, and displacement of populations. In the long term, the asymmetrical power of the French and their counter-insurgency tactics were effective, since sympathy for the rebels gradually declined. When the revolt was finally crushed, 6,000 rebels had been killed and more than 100,000 civilians were left homeless. By 1927, the rebels were not welcome in most urban areas, and the population was afraid of possible French retaliation if the rebels were allowed to operate in their region.[27]

After the defeat of the rebels, the National Bloc appeared as the main political force. The coalition was made mainly of landowners from the four main cities. They were 90 percent Sunnis and had a reconciliatory approach to politics. The National Bloc was the main nationalist force after the defeat of the 1925 Revolt. Writing about African countries, Nigel Gibson explains that this type of nationalism produces a native bourgeoisie that depends on Western capital for its survival.[28] He notes that this:

> ..."caste" [is] essentially an unproductive caricature of the Western bourgeoisie, that then assumes national leadership. In this scenario, independence does not lead to decolonization but to a curious self-recolonization where a native leadership simply mimics the privileges and postures of the Europeans and follows it on the path towards "decadence" [...] while the masses sink deeper in poverty.[29]

Popular nationalism of the early twentieth century gradually metastasized to become an instrumentalist nationalism by mid-century. It was a reconciliatory nationalism that often avoided confrontation with the colonial power.

ARAB NATIONALISM AFTER INDEPENDENCE

The second pivotal moment in the history of nationalism in Syria starts in the early 1960s. It revolves around the Baath party's authoritarian rule since 1963 and the despotic ideology it produced. The history of the Baath party and its rise to power is beyond the scope of this chapter, but suffice to say that the period from 1946, when Syria became independent, to 1963, when the Baath party seized power, was a turbulent one. Nationalist, communist, and Islamist groups were competing for power, sometimes using democratic means, while at others times exercising force. This period could be described as nationalist *par excellence*. Due to structural transformations, the notables began losing their economic and political power. The notables' politics became ineffective, and notables had only two options: either become marginal, or join nationalist parties.

During that period, in addition to the notables' politics, the Baath and Communist parties dominated the political scene in Syria. Interestingly, there are several parallels between the rebels of the Great Syrian Revolt and the intellectuals of the Baath party. Salah al-Din al-Bitar and Michel Aflaq were both the sons of grain merchants from the al-Maydan Quarter in Damascus, and their ally in the rural south was the son of Sultan al-Atrash, the leader of the Great Syrian Revolt.[30] The Baath party utilized the register of the Great Syrian Revolt to gain legitimacy. Many members of the Baath were teachers and doctors who knew the history of the 1925 revolt, and sometimes emulated its strategy by building lasting relationships between the city and countryside. They did so by opening clinics in remote villages and educating the sons of peasants. However, the parallels between the 1920s and 1950s should not be pushed further. The openness and inclusiveness of the Baath gradually disappeared as the party became hegemonic.

The intellectuals of the Baath party produced a very rigid discourse about Arab identity that led to an exclusive brand of nationalism. In that regard, the Baath nationalism was unlike that of the 1920s. Nationalist leaders of the 1920s built a popular nationalism, which was vital for self-determination struggles, while those of the 1950s were primarily interested in seizing and maintaining state power. In the first case, nationalism was inclusive, organic, and anti-colonial, while in the second, it paved the way for an authoritarian, exclusive, and neo-colonial ideology. The former created a nationalism that could have produced an independent post-colonial state, while the latter used the state to create an instrumentalist nationalist rhetoric, the main goal of which was to produce legitimacy for the ruling party and its leaders. It is no coincidence that, between 1963 and 1970, civilians within the party were marginalized, while Baathists with military backgrounds gradually took over.

IS THE CURRENT REVOLT DECOLONIAL?

The current revolt is producing its own nationalism, while at the same time countering the despotic ideology of the Baath party. As such, two nationalisms are competing for dominance. The first one is popular and anti-dictatorial, while the second is official and claims to be anti-imperialist and anti-Zionist. The first resonates with the nationalism of the Great Syrian Revolt of the 1920s, while the second is the heir of pan-Arabism of the 1950s and 1960s. The first was experimental, while the second followed a scripted program. The former's main objective was decolonization; the latter's goal was, and remains, autocratic rule. Nationalism of the 1950s and 1960s was influenced by Nasser's centralization of power, but had a much less ambitious developmentalist goal.

The current uprising is producing a new nationalism that strives to delink gradually from the state-centric nationalism of the Baath party and reinvent itself, in part, by reconnecting with the nationalism of the early twentieth century. The process of delinking is delicate and risky. If it is performed too slowly, it could be co-opted or subverted. However, the rejection of state-centric nationalism, while desirable, could leave a discursive space that sectarian or despotic groups could occupy. This explains why supra-nationalist and infra-nationalist groups with Islamic, tribal, and ethnic identities dominate the scene. Due to these competing identities, which are functioning at the local and trans-local levels, popular nationalism is facing a major challenge as it attempts to counter state nationalism.

Popular nationalism, which this chapter investigates in more depth in Manbij, should be understood as a dynamic cultural praxis. It could morph into a Manichean ideology; it could also evolve into a Fanonian new humanism. In addition, it could be overtaken by other competing ideologies. Most opposition leaders are detached from people's everyday realities. They are mostly busy producing centralized and exclusive narratives that are in many ways a replica of the regime's ideology.[31] The nationalism that protesters are producing with their bodies, emotions, dances, and various praxes is promising, but has been gradually marginalized.[32] Popular nationalism was dominant before the militarization of the revolution, and stayed influential until the end of 2013. It is also competing with transnational Islamism in the liberated regions such as Manbij.

In the following section, popular nationalism is explored in the context of the liberated city of Manbij between 2011 and early 2014. In those liberated regions, nationalism is deployed to solve everyday problems. As Fanon explains: "The living expression of the nation is the moving consciousness of the whole of the people; it is the coherent, enlightened action of men and women."³³ Against the Manichean worlds that both the regime and a sizable segment of opposition groups are attempting to impose on Syrians, people are resisting by producing polyphonic narratives. Unlike official nationalism, decolonial nationalism is plural. It is based on a politics of rupture from the nationalism of the 1960s and a critical rediscovery and reconnection with the nationalism of the 1920s.

MANBIJ: A CASE STUDY

The study of popular nationalism in contemporary Syria necessitates an exploration of everyday praxes, beings, and spaces to comprehend its scope and implications. It is the result of myriad experimentations that Syrians are undertaking in various sites. Unlike official nationalism, which is centralized, rigidly structured, and unambiguous, popular nationalism is decentralized and multi-vocal. It develops through countless interactions in numerous localities. Baath nationalism is based on unchanging abstract narratives, while popular nationalism is iterative and adjusts constantly to adapt to new situations. Popular nationalism is adaptive and highly malleable, since it is still in its early period of formation. It is saturated with uncertainties.

To explore the significance of the emergent paradigm, the next section focuses on Manbij, a strategic space for the study of the various moments of popular nationalism. These moments are not necessarily generalizable to other regions, but could help us comprehend the complexities of an emergent identity. The northern city, which was controlled by the revolutionaries for 18 months from mid-2012 to late 2013, provides an excellent point of entry for the exploration of three vital moments. The first takes place during the early period of the revolt, before the liberation of Manbij. This period is characterized by subterranean nationalism that attempts to find its way to the surface. It is a clandestine nationalism that could be explored in marginal localities and peripheral spaces. The prison is one such paradigmatic space, where activists and organizers created a new imagined community during the early phase of the revolt. The second phase takes place after the liberation of the city in July 2012. It consists of an urban confrontation between the citizens of Manbij and several Salafist groups, who tried to hijack the revolt and impose their sectarian vision of Islam. The main terrain of contestation is the public space and the ways it was appropriated by hegemonic Islamist groups. The third phase began at the end of 2013, when the Islamic State in Iraq and Syria started a campaign of terror in Manbij. A group of activists created an organization they named "Home" to protect the revolution and preserve the nascent nationalism. In reality, it was a semi-public space that allowed them to operate in the city, despite the ISIS presence.

CLANDESTINE NATIONALISM

The emergence of popular nationalism in Syria should be conceptualized as an iterative process that adapts to various environments. It is the antithesis of official nationalism, which is produced by political leaders who define the contours of the political community and the profile of its citizens. Popular nationalism is a practical nationalism that solves everyday problems. It is built through people's struggle and their resistance, but also their aspiration for a better tomorrow. Early nationalism that emerged in Manbij and other cities was clandestine. In the early days, it was attempting to find its way to a large audience under challenging conditions. This is why it developed in dangerous or marginal spaces and was used to resist the regime's violence. Like popular nationalism of the early twentieth century, which emerged as a tool of resistance against French hegemony and colonization, the nationalism that began to emerge in 2011 should be understood as a vital tool for opposing the regime's violence. To explore this form of nationalism, it is necessary to analyze the work of clandestine committees that appeared in many cities, including Manbij. Their goal was to organize protests and other actions to undermine the power of the security apparatuses and the army.

Several such committees were created, modeled to a certain extent after the Egyptian experience of neighborhood committees in Cairo, Alexandria, and other Egyptian cities. The main goal of these committees was to prepare for autonomy under challenging conditions. The committees were usually composed of individuals willing to take high risks to liberate their cities. Like their predecessors in the 1920s, they had a deep conviction that struggle against the Assad regime was primarily about decolonizing the authoritarian spaces in which they were forced to live for several decades. The politics of dignity played a central role in the emergent revolutionary culture. In mid-February 2011, during the first protest in Damascus, the crowd chanted, "The Syrian people will not be humiliated!"[34] The new nationalism posited itself against authoritarianism, terror, and humiliation. Syrians were committed to liberating their city peacefully through popular protests. Gradually, however, it became clear that the violence of the regime would only increase, and peaceful struggle would not be enough.

After the prayer in the Grand Mosque of Manbij on April 22, 2012, activists held their first public protest.[35] It was attended by thousands of residents, who chanted against Assad and his murderous war against Syrians. A few weeks earlier, the Syrian Army had committed a massacre in Taftanaz[36] city center and Karm al-Zeitoun in Homs.[37] Many residents had joined the Free Syrian Army and were fighting on different fronts, but those who stayed in Manbij were determined to liberate their city.

A secret meeting was organized, and the attendees decided to form several neighborhood committees (North, South, East, and West), where coordinators in each committee would recruit individuals in their respective neighborhoods. The revolutionaries created a structure designed to prevent or limit infiltrations by regime informants. They contacted trusted intellectuals and individuals, and began building networks that functioned almost independently from each other in various neighborhoods. Activists followed the news of revolts in other countries very closely, and once the protests reached Dara'a, the first city to witness mass organizing against the Syrian regime, the residents of Manbij began thinking about ways to support Dara'a and other rebellious cities, while at the same time expelling regime forces from Manbij. The discussions revolved around developing strategies to counter the violence of the regime. A radical transformation was taking place in Syrians' mode of thinking. It was the equivalent of a psychic cataclysm in their collective imaginary. The primary purpose of the insurrection was people's need to regain their humanity and dignity. People quickly realized that spaces of encounter are vital for creating democratic political alternatives. In that regard, popular nationalism cannot be understood outside the politics of dignity and the myriad stories of struggle against daily humiliation and injustice.

During this early period, Syrians experienced a form of decolonization of bodies and beings. It was the result of open discussions about the history of Syria and current political conjuncture. They started developing strategies to fight against the injustice that their people had endured for decades. As noted above, Fanon explains that national liberation in the Global South is incomplete without a radical transformation of culture and economy. That is the main reason why liberation in these countries was pre-empted by neocolonial policies. After independence, the bourgeois class in many African and Asian countries became a comprador class that depended on the West to maintain its grip on power. Their role was instrumental in crushing the aspirations of popular classes. It is in this context that national liberation in Syria should be understood. Revolutionaries were able to create decolonial spaces for the popular classes in Manbij to discuss the future of their city and link their destiny to the struggles of other Syrians who were organizing in Dara'a, Homs, Hama, and elsewhere.

Revolutionaries organized various actions at night, many of which should be read as part of the construction of a political community and an emergent national identity. After a period of several months of intense discussions in clandestine spaces, they took their struggles to the street, knowing they were taking high risks by doing so. Their actions consisted of targeting the symbols of official nationalism embodied in the pictures of Assad and Baath slogans that saturated the public space. They filled the walls of the city with graffiti against the symbols of the decaying regime. The main task for revolutionaries was to erase the traces of the old regime, and as such to decolonize space and being. Unlike official nationalism, which aspires to maintain the structures of oppression, the embryonic nationalism in Manbij was deployed to undo them. As such, it cannot be delinked from the politics of dignity that one could witness in the streets and squares of Manbij, and in which the Syrian revolution is rooted.

Despite their importance, most of these early actions were nocturnal.[38] The inhabitants of Manbij started protesting in the streets, drawing graffiti on the moribund walls of their city, tearing down pictures, disfiguring statues of the Syrian despot and his father, and liberating their neighborhoods from the oppressive symbols of the old order by erasing them or painting over them.[39] The revolutionaries documented these attacks by taking pictures and making videos, and posting them online.[40] By doing so, they were inserting their struggles into a larger national narrative of resistance. There were a multitude of nocturnal actions, many of which took place independently from each other. Activists were building a new culture through their nocturnal journeys and online organizing, and by doing so, they were subverting the official narrative about Manbij.

When the revolution began, most residents believed nothing could happen in Manbij, because the city was perceived as too loyal to the Assad regime.[41] These fragmented struggles were reassembled online and given a more coherent meaning. This iterative process between nocturnal actions and online organizing built momentum in Manbij. Many residents organized small-scale actions and posted them online without revealing their identities.[42] In some videos, activists covered their heads, while in others, they were seen from behind. The emergent online community undermined the narrative sustained by the regime's representatives in Manbij.[43]

Clandestine nationalism was taking place in another important location during this early phase of the revolt. Many of the activists in Manbij and other cities were arrested and put in jail in Aleppo and Idlib, while others were sent to Damascus. Prison became an important space of encounter. For activists, it symbolized dictatorship and oppression in post-colonial Syria. Popular nationalism is, in many ways, the antithesis of what Syria has become under Assad's rule: a large prison. Since the beginning of the revolt, the regime had arrested a large number of activists from rebellious cities to interrogate and torture them, and to terrorize the population. This strategy backfired, as revolutionaries used their time in prison to meet one another and share their stories of resistance and organizing against dictatorship. In the early phase of the revolt, activists risked their lives by traveling from one city to another to learn from the experiences of others. Ironically, by incarcerating them, the regime created the ideal space for activists. It mixed people from different regions in the prisons of Damascus, Aleppo, and other large cities, and as such provided a space where activists were able to share subversive knowledge and plans at the national level. Fanon writes,

> The Algerian national culture takes form and shape during the fight, in prison, facing the guillotine [...] national culture is no folklore where an abstract populism is convinced it has uncovered the popular truth. It is not some congealed mass of noble gestures, in other words less and less connected with the reality of the people. National culture is the collective thought process of people to describe, justify, and extol the actions whereby they have joined forces and remained strong.[44]

This type of space was ideal for the construction of popular nationalism. Syrians who had never had the opportunity to live in such discursive spaces were now inhabiting them. In the early days, the regime was unable to keep activists for more than a few weeks, since it had limited space. During that period, the regime did not kill activists in prison in a systematic way, thinking that such actions would backfire and amplify the scale of the revolt. This obviously does not absolve the regime of the violence it enacted on activists. Activists were released from prison after a few months not because they were a lesser threat to the regime, but because security branches needed to free space for other protesters. By doing so, the revolutionaries of Manbij who had spent time in prison during this early period shared their stories with others, and learned from them. Together, they created a common ground for ongoing and future struggles. Many activists from Manbij tell stories about the formation of a national identity in prison, articulated around the concepts of resistance, dignity, and justice. Despite the atrocities, prison was a space where activists met others from various regions and different backgrounds, and together discussed the meaning of their struggle at the national level. Their conversation showed that a new national identity was necessary, but could not be built on the old foundations of the Baath nationalism and its hollow slogans. What characterizes nationalism of the first stage is its performative nature and its ability to adapt quickly to the needs of everyday struggles. This nationalism was taking place in peripheral spaces such as the prison. It was also performed at night in central squares that symbolized the coercive power of the central state.

FROM BREAKING THE SILENCE TO DECOLONIZING SPACE

The second phase of popular nationalism erupted with the liberation of the city. The process was gradual as protesters occupied more central and visible spaces. Unlike the first period, their actions were now taking place during the day. An increasing number of residents moved their rebellion from the intimate spaces of the household to the street and the neighborhood. They turned the minutiae of everyday resistance into a fully-fledged and mature revolt. Popular nationalism was not confined to the nocturnal hours anymore. Nationalism was an important tool for activists to decolonize their neighborhoods and prepare for the real liberation of Manbij, which requires more than expelling the security forces and state representatives from the city.

Discussing the Arab revolts, Agathangelou explains that the youth created a new relationship between bodies and space, which allowed them to produce a revolutionary political community. She writes, "bodies became the sites that contested those objects and technologies that systematically worked to segregate, discipline, contain, kill them, or let them die."[45]

Unlike official nationalism, which has a long intellectual history and is mostly based on the writings of Baath intellectuals, recent popular nationalism is not scripted. One of the important sites for its exploration is the urban texture. In this section, popular nationalism is analyzed through the spatial reconfiguration of the city. The destruction of the Baath party's symbols and their replacement with ones that embody the values of the revolution was an integral part of this phase.

The primary goal for protesters in Manbij was to regain their humanity and uproot "the sense of nobodiness"[46] from their minds and bodies. Several residents explained that by challenging a despotic order, the uprising allowed them to regain their dignity. Fanon writes, "[t]here is a zone of nonbeing, an extraordinarily sterile and arid region, an utterly naked declivity where an authentic upheaval can be born."[47] It is against this zone of nonbeing that the insurrection in Manbij should be posited. The rebellious population created spaces where one could become a human being again. The slogan *"Syrian people will not be humiliated!"* was a loud cry of freedom from within the zone of nonbeing. In this context, dignity should not be understood as simply a slogan to arouse the masses against the regime. It signals a shift in the theoretical, epistemic, social, and political spheres that could lead to radical transformations in Syrian society. American intellectual John Holloway notes,

> [t]he whole relation between theory and practice is thrown into question: theory can no longer be seen as being brought from outside, but is obviously the product of everyday practice. And dignity takes the place of imperialism as the starting point of theoretical reflection.[48]

The Syrian uprising, and the Arab revolutions more generally, allowed people to negotiate dominant political ideologies whether progressive or reactionary. The politics of dignity embodied in the chants, slogans, and graffiti filled the streets of Manbij after the events of April 2012. These symbols were meant to subvert spaces controlled by the regime. They were the nucleus of a nationalism that is inclusive and emancipatory. In this context, popular nationalism was produced by bodies in movement that gradually decolonized the streets of Manbij from the hegemonic power of the Baath party and its ossified nationalism. In July 2012, the city of Manbij was finally liberated after several months of peaceful protests.

Dignity should guide our analysis in the vast political labyrinth of the Syrian revolution. In this context, dignity should be understood as a hymn of national liberation as it provides Syrians with the conceptual apparatus and performative practices to move from a zone of nonbeing to multiple spaces of existence. The echoes of the slogans in the streets of Damascus and Dara'a represent more than protests against an oppressive regime: they are part of an ontological revolution and a national liberation. Residents in Manbij explained that hearing their own voices in the street, for the first time, was a transformative experience; repeating, *"Syrian people will not be humiliated!"* or *"The people want the fall of the regime!"* is in itself a liberatory performance. For protesters, dignity embodies an aspiration for a democratic nation.

Fanon explains that during the Algerian revolution, decolonization of body and mind, the process of regaining a dignified life, was at the core of the struggle, and preceded the independence of a nation colonized by an imperial power. The liberation of territories from security and military apparatuses is a main pillar of popular nationalism. As such, popular nationalism cannot be understood without positioning bodies of protesters at the center of the analysis.

Popular nationalism in Manbij can be examined through the minutiae of people's everyday struggles, the movement of their bodies in the public space, and the reverberation of their voices in the street. For an external observer, the Syrian uprising seems like a cacophony of voices lacking structure or purpose. A close examination, however, reveals an effervescent society struggling not only for national liberation, but also for new foundations of culture and history, the meaning of the nation and being. Every aspect of Syrian society is experiencing a radical transformation. In Manbij, an increasing number of residents were preparing their city for liberation in spring 2012. A revolution was taking place at every level, starting from the mundane aspects of everyday life to the most complex facets of society, culture, and nation.

After the liberation of Manbij, revolutionaries removed the symbols of the old order and replaced them with graffiti that represented the diverse cultures of the city. An analysis of public space in Manbij could help us to understand the construction of the new imagined political community. As indicated above, popular nationalism since 2011 has been operating a dual move: 1) a connection to a past that precedes Assad's rise to power; and 2) a rejection of official nationalism. After the liberation of the city, Syrians erased all symbols of Assad and his regime by removing his statues and pictures. They replaced them with graffiti representing the martyrs of the uprising and revolutionary slogans. They renamed certain streets and public squares, and beautified the walls in the city center.⁴⁹ Near the revolutionary council, activists painted a portrait of Bassel Shehada, a Christian film-maker who left the United States, where he was studying, to return to Syria and participate in the revolution.⁵⁰ Beneath his picture was one of Mashaal Tammo, the Kurdish leader who was killed by Syrian security in Kamishlo in October 2011.⁵¹ Beneath them was a drawing of Hamza al-Khatib, the 13-year-old boy who was tortured to death by Syrian security in May 2011.⁵² Several slogans about democracy and inclusion accompanied the graffiti. One of them stated, "[s]ectarianism is a bullet that will kill the revolution."⁵³

However, the patriarchal culture in Manbij invisiblized women and their struggles. Revolutionary women were not depicted on the walls of the city. This is tragic, since women have played a central role in the Syrian revolt since the beginning.⁵⁴ In Manbij, women organized several important campaigns through the Future's Youth and other organizations.⁵⁵

In the months following the liberation of the city, newly formed groups initiated numerous projects to further the decolonization of their city. Several were proposed by "Future's Youth organization," a group of activists whose goal was to improve the living conditions in the city.⁵⁶ They organized campaigns to help internally displaced persons, educate children and keep them away from the streets, and provide equipment and medication to hospitals. Just as importantly, the group planned a number of symbolic activities to clean the streets and walls of Manbij. They recruited young residents for these campaigns to help transform the city. One of these campaigns was named, "My City is My Home," and was meant to change the residents' relationship with their city.⁵⁷ It invited the residents to take back their city and remove anything that reminded them of the criminal regime.

The cleaning campaign had a dual meaning: the first is literal, the second symbolic. First, by removing the piled trash from the sidewalk, the residents produced a more hygienic and enjoyable city. Second and more importantly, they needed to cleanse Manbij of all symbols of the old order. To help achieve that goal, they organized a campaign to beautify the walls. The Future's Youths invited artists to paint revolutionary graffiti and slogans on the walls near the city center. A group member explains, "[We] are decorating the walls in order to liberate them from the regime's grip."⁵⁸ One piece of graffiti depicted the corrupt neoliberal class that benefited from Assad's rule. It was represented as a wealthy man devouring a pile of food by himself, without sharing with anyone. Another one showed a colorful mural of Manbij, with an inscription: "A happy announcement to the residents: there is an explosive barrel bomb for every house, every school, and every bakery."

In March 2013, activists in several cities, including Manbij, organized a street festival to celebrate the second anniversary of the revolution. In addition to film screenings and a theater performance, several outdoor workshops about beautifying the city were organized.[59] These performances were very well attended although women were mostly absent. The flag of the revolution could be seen throughout the city. It was the symbol of a new Syria and a future without Assad. This festival and other cultural activities were meant to delink Manbij from its despotic past and give people a chance to re-appropriate their city.[60]

After the liberation of the city, activists organized several campaigns to paint the flag of the revolution in the city and remove the remaining symbols of the Baath party. These activities were meant to decolonize the spaces that were previously occupied by regime forces. The organization of concerts in the city center was part of that re-appropriation of space that the Baath regime had sullied. In addition, the juxtaposition of contradictory styles of music, namely, rap and traditional, was symptomatic of the new Syria and its emergent popular nationalism. The rebellious rap music signaled a clear rejection of the old totalitarian order, while traditional music showed an attachment to a past untainted by Assad's rule.

The flag of the revolution, which emerged in 2012, represented a clear rejection of the old flag, which the Assad regime adopted in 1980. The revolutionary flag was from the post-independence period, but was replaced by the Baath party in 1963.[61]

After the liberation of the city, the central prison, an important symbol of Assad's tyrannical order, was turned into a local attraction. People who visited the carceral space saw dark and filthy micro-cells and were reminded of Assad regime's history of violence. The inhabitants of Manbij were persuaded that the old order belonged to a morbid past, and that a new era without Assad was possible.

By the end of 2012 and early 2013, several military factions tried to control the grain mills and bakeries under the pretext that bread is a strategic commodity, and as such should be protected by powerful military groups. Al-Nusra, an offshoot of al-Qaeda in Syria, the Islamic State in Iraq and Sham (ISIS), and Ahrar al-Sham, a large Salafi group present in the city of Manbij, all tried to take control of the mills, silos, and bakeries;

but the revolutionary council mobilized the population and organized a robust resistance to prevent them from taking over these vital institutions. Part of the multifaceted resistance against Islamist groups took the form of graffiti and slogans on the walls of the city. During this period, slogans such as "ISIS Out!" and "Ahrar Out!" could be seen in different neighborhoods, and especially the city center. Before the jihadist groups' presence in the city, the residents organized a protest on September 28, 2012 to call for the unification of all FSA groups in the city. When ISIS tried to remove the flags of the revolution, kidnap activists or religious clerics who disagreed with the terrorist group's politics, or take over the revolutionary council, they faced immense popular resistance.

For example, on July 9, 2013, city residents and activists organized a protest in front of ISIS's headquarters. The protest was also one of the first in Syria, and signaled to the terrorist group that massive resistance to their takeover should to be expected. Protesters chanted slogans against the Islamic State: "ISIS Out! Manbij is Free!" and "Your state is no different from Assad's!"[62] From that point on, the relationship between the revolutionary council and ISIS became very tense. The president of the council received multiple death threats and was forced to work from home. ISIS began removing the flags of the revolution and replacing them with their black flags. The jihadist group wrote inscriptions on the city center's walls that reminded residents that ISIS's real enemy is not the Syrian regime, but revolutionary forces. These slogans were against secularism, nationalism, freedom, and democracy.[63]

Most residents perceived these Islamic groups as an extension of Assad's regime. Their efforts to suppress grassroots movements and impose their own ideology was similar to the regime's despotism in many ways. Popular nationalism in Manbij was not posited against Islam. Rather, residents were opposed to the ISIS brand of Islam. Most people rejected their sectarian and fundamentalist ideology and viewed them as foreign invaders. One activist who worked with the revolutionary council reminded people in summer 2013 at a protest against the terrorist group, "We are Muslim! We don't need anyone to Islamize us." Popular nationalism during that period was pushed back to the private and semi-private space.

INCUBATOR NATIONALISM

After the protests against ISIS, popular nationalism took a different and subtler form. It receded into a semi-private space, as its appearances in the public arena had become riskier. Black paint gradually covered revolutionary slogans, while the flag of the revolution was replaced with ISIS's black flag. The best representation of this period is probably the establishment of "Home," a youth organization that created a communal space for the residents of Manbij in the midst of the gradual control of their city by ISIS fighters. The young activists conceived the project as a space of resistance against ISIS. Home was established in September 2013 to work with children and teenagers on several projects. It screened movies and offered painting workshops to children, and Turkish classes to residents who planned to move to Turkey. Paradoxically, Home's organizers were helping build a homeland abroad by teaching residents Turkish and assisting them in leaving the city. Most importantly, Home opened a space where the youth of Manbij could discuss arts, politics, the revolution, and the fate of their city, and by extension their nation. It was named "Home" because its founders felt that the revolutionary spaces were shrinking in Manbij, and that the situation in Syria was becoming more complex. One of the activists involved in the project explains:

> The Syria we were dreaming of, and aspiring to create, is becoming an illusion. Sectarianism and fundamentalism are eating us up. This is why we felt the need to create a home to preserve whatever can be preserved and start planning all over again.[64]

The campaign named "My City is My Home," which was described above, was paradigmatic of that second period, while the youth project "Home" became the focal point during the third period. Despite their lack of funding and the repeated threats by ISIS, they operated for several months under considerable pressure. ISIS understood the danger of leaving such a space available, knowing it would function as an incubator for nationalism and Syrianness, both of which were antithetical to ISIS's transnational sectarian ideology. During this transitory period, Manbij became the hub for ISIS foreign fighters. The city was dubbed "Little London" due to the large number of British jihadists who lived there. In addition, French, Germans, Chechens, Russians, and other foreigners moved into Manbij to participate in the rebirth of the Islamic caliphate.[65] Evidently, these ISIS foreign fighters considered Syrian nationalism a major threat. For them, nationalism was a secular Western ideology that had to be crushed at any cost.

In January 2014, clashes between opposition groups and ISIS erupted in most of Northern Syria. Revolutionaries evicted ISIS from Manbij and prevented the jihadist group from re-entering their city for almost two weeks. But ISIS came back with reinforcements, and shelled Manbij for several days before taking it back. The founders of "Home" fled the city and put an end to an important initiative. Some of them are currently living in Turkey, where they continue their tumultuous journey by protecting what remains of Syrian popular nationalism.

CONCLUSION

The three phases of popular nationalism, namely, clandestine, decolonial, and incubator nationalisms, show that the emergent discourse in Syria is facing major challenges as it develops and adapts. The nationalism that emerged in Manbij was primarily meant to counter the Syrian regime's authoritarian discourses. Popular nationalism constitutes a culture of resistance that was gradually delinking from the oppressive structures of the official nationalism of the Baath party, while at the same time reconnecting with earlier histories. The popular nationalism that developed in the early twentieth century to end French colonization and the one that has emerged since 2011 as an oppositional paradigm to counter "Suryia al-Assad" (Assad's Syria) have much in common. Both produced non-elitist and non-scripted cultural paradigms. Both developed iteratively through experimentation to address everyday problems in a context of war and population displacements. They are oppositional to the state-centric nationalism that was rejected by the population because of its oppressive nature.

Popular nationalism is at a crossroads, since it has been marginalized and challenged by state and non-state actors alike. Popular nationalism in Syria is not only facing the nationalism of the Baath party, it is also competing with Muslim identities imposed by groups such as ISIS, Ahrar al-Sham, and Jabhat al-Nusra. In addition, it was challenged by sectarian and tribal identities operating in multiple localities. Finally, it was also being redefined by Kurdish nationalism, which has become dominant in Northern Syria and in Manbij since 2016. Some of these identities are operating below the nation-state, while others are functioning above it. They are challenging popular nationalism to be more creative, open, and inclusive. This is one of the major challenges for popular nationalism in the coming years, as the Arab revolts transform the landscape in the region.

4
The Politics of Bread and Micropolitical Resistance

> *The Syrian people are not hungry*
> *The People want the fall of the regime*
> Syrian protest chants

This chapter examines the politics of bread in Syria since independence. It starts with a history of the agrarian reform that Abdel Nasser implemented in Syria in the late 1950s as well as the policies put in place by the Baath party since 1963. It explores the Syrian regime's strategy of curtailing political rights while providing inexpensive bread. The first section explores the agrarian reform and the technologies of population management. Prior to the revolt, the state used bread as a tool to control and discipline the population, and to regulate the lives of its citizens. Since 2011, it deployed new strategies to build loyalty among the residents in the areas it controls, and to cripple the communities living in the liberated regions. The chapter shows that the Syrian state used bread as a lethal weapon during the revolt to suppress opposition and undermine revolutionary processes. The regime bombed bread lines to break the revolutionaries' morale and burnt wheat and barley fields to starve the liberated areas.

The last section maps the tactics utilized by revolutionaries in Manbij as they build and maintain an autonomous economy of bread. The chapter argues that state power should not be reduced to the coercive power of the military and security apparatuses. The Baath party built a robust assemblage in rural areas that allowed it to control every aspect of peasants' lives. French philosopher, Gilles Deleuze proposed the concept of the assemblage as a way to understand the complex matrix of power. He explains that state power is constituted of networks dispersed throughout a territory that helps control it. It constantly captures new spaces by expanding its perimeter and thus making it virtually impossible to escape its grip. In the Syrian context, the Baath used the agrarian reform, the construction of dams, and the creation of irrigation networks, among other tools to impose an assemblage to control the lives and subjectivities of the rural population. The task of the Baath assemblage was to capture any spaces previously outside its purview. The Baathist state made it gradually more difficult for peasants to have any form of autonomy in the rural areas.[1]

LAND REFORM

While the French Mandate built a solid alliance with powerful landowners, their relationship was conflictual during World War II. The colonial power gave landowners numerous privileges including tax exemption on farming. Following in the footsteps of the Ottomans, French bureaucrats believed that powerful landowners would be useful in helping quell peasants' protests and protect France's political and economic interests. When mechanization reached the Syrian countryside, it produced a greater concentration of wealth in the hands of powerful landowners, and immiseration of large segments of the peasantry, reducing many to only seasonal work. During World War II, the colonial power had a tense relationship with landowners and peasants for their refusal to surrender their grain production at a price below market value. French bureaucrats were also facing frequent unrest in large cities. The urban population was dissatisfied for two reasons, namely, low wages and the high price of bread. Populist politician Shaykh Taj al-Din, appointed by France as president of Syria, was tasked with collecting funds for the war effort. He tried to collect taxes from the middle classes to subsidize bread while also driving up the national debt. Then, to address the mounting debt, he raised bread prices and allowed merchants in Aleppo and Damascus to profit from the war by letting them sell their grains at an inflated price.[2] The anti-colonial movement gained momentum by organizing hunger marches against the French Mandate and the government of Shaykh Taj al-Din. Political scientist, Steven Heydemann notes that nationalist notables, "took over leadership of hunger marches from unions, women, and others and coordinated a sustained campaign against the alleged corruption of Prime Minister Barazi, ousting him in December 1942."[3]

After independence, wealthy landowners and merchants dominated political life in parliament, while communists and Arab nationalists parties, namely, the Baath and the Arab Socialist Party, headed by Akram al-Hourani, were gaining traction in the street. Land reform was one of the mobilizing issues for the base. Hourani was an excellent orator and a central figure in Syrian politics in the post-independence era. He created a party with a radical agrarian program, by advocating for land distribution and the end of feudalism. Palestinian historian, Hanna Batatu notes that Hourani, who created a party against landlords in 1943, popularized the slogan, "Fetch the Basket and Shovel to Bury the Agha and the Bey," to highlight his disdain for the landed oligarchy. He made the struggle against the feudal class a focal point in his political strategy. He believed the only way to improve peasants' conditions, and end feudalism, was to enact a profound agrarian reform. Batatu adds, "Hawrani returned from Palestine convinced that 'feudalism' was the cause of the Arab defeat and that the agrarian question and the Arab national cause were closely linked."[4] Hourani had a lasting impact on Syrian politics and more specifically on the peasant question. After his trip to Palestine, he became persuaded that land reform was essential, not only to end peasants' plight, but also to build a modern nation capable of developing its economy, and defeating Western imperialism and Israeli Zionism.

Land reform was a polarizing topic among the Syrian ruling classes, who had divergent interests. After independence, the industrial bourgeoisie began encroaching on the power of the landed oligarchy. To consolidate their economic power, they needed to create a national market and a new middle class with the means to consume locally manufactured products. In the early 1950s, 55–60 percent of the population was composed of peasants and the vast majority were extremely poor.[5] Syria's industrialist class was mainly based in Aleppo and used the People's Party, a moderate center-right party to further its interests. The People's Party represented principally the Aleppan ruling class in parliament and often clashed with the National Party, which was based in Damascus. Aleppo was the wealthiest and largest city in Syria until its gradual marginalization after World War I. Since the Sykes–Picot Agreement and the creation of borders between Iraq and Syria, Aleppo suffered tremendously from the separation of its Iraqi trading partners. By losing its long-standing trading networks with Iraqi cities, Aleppo's economy gradually declined. The People's Party wanted to revive these trading networks by building a union with Iraq, while the Nationalist Party opposed it vehemently.[6]

In 1950, the People's Party, with the support of nationalist forces in parliament, passed a new constitution with provisions regarding land distribution and limits on lot size.[7] The People's Party, which represented the interests of industrialists, considered the distribution of state land to landless peasants vital for a new middle class with the means to consume local production to emerge. However, the reform from above failed because nationalist and communist forces were on the rise, and were pushing for a more radical reform that the landed oligarchy and their allies could not accept. After independence, new nationalist forces, such as the Baath party forged an alliance with the People's Party, and advocated for an import substitution economy and land reform. The National Bloc (which later became the National Party) perceived land reform as a direct threat to its economic interests, and opposed it fiercely in parliament.[8]

The alliance between nationalist forces and the People's Party fell apart in 1953–1954. In the early 1950s, Syrians were drawn to Arab nationalism and socialism due to several factors, among these the Palestinian Nakkba, the 1949 CIA-engineered coup in Syria, and the Soviet support for progressive causes in the region. As a result, people demanded more radical changes that traditionalist leaders could not sanction. The new political environment forced traditionalist parties to overcome their differences and build a conservative alliance against Baathists and Communists. To prevent the complete destruction of the dominant social order, and the takeover by radical forces, the ruling classes (industrialists, merchants, and landowners) were forced to work together, despite their divergences. In the mid-1950s, after failing to implement a moderate reform from above, with the help of the Baath and other radical parties, the People's Party forged an alliance with its arch-enemy, the Nationalist Party. The collapse of the reformist alliance opened the door for more radical possibilities.[9]

Heydemann notes,

> In Syria as in other countries, a conservative coalition between the declining landed elite and an emerging bourgeoisie emerged to confront growing mobilization among peasants and workers. Land and capital joined to defeat an increasingly militant force of workers and peasants. Efforts by capitalists to achieve Syria's industrialization by means of controlled, liberal social pact had failed.[10]

Unlike European countries during the eighteenth and nineteenth century, where a strong industrialist class built a robust coalition and annihilated the feudal system, Syria's industrial bourgeoisie was unable to impose its economic program on traditional forces.

In the late 1950s, the Syrian Communist Party (SCP) pursued its ascendency by mobilizing workers and peasants in the street and becoming a formidable force in Syrian politics. In 1957–1958, traditional parties and the Baath viewed the union with Egypt as a better option than a power takeover by radical forces, especially the SCP. Communists were not always popular in Syria especially during the 1940s. In 1947, Moscow accepted the partition of Palestine, which became the official position of many communist parties worldwide, including the SCP. Thousands of members left it to join nationalist forces, including Akram al-Hourani's party. In the late 1950s, the SCP regained a certain momentum due to domestic and international factors, and was perceived as a threat by Baathists and traditional parties. The Union with Egypt was the best option for these forces to prevent Communists' rise to power. During the Union with Syria, Nasser banned all political parties, which led to the gradual decline of communism in Syria. When the SCP re-entered the political arena in 1966, it had already become insignificant.[11]

During the Union, Nasser built an alliance with Syrian capitalist classes, who were essential for the success of the developmentalist project. He was determined to destroy the feudal class, and did so by implementing a comprehensive agrarian reform. His goal was to end industrialists' dependence on the landed oligarchy, which in turn would help him build the industrial sector in Syria. Peasants' work was regulated by new laws, including a minimum wage and better working conditions. In addition, Nasser required that all peasants join a labor union by 1960. Law 143 regulated the relationship between landowners and peasants and systematized collective bargaining, while it made sure to address the injustice inherited from the feudal system. Members of the peasants union were prohibited from engaging in any activities that could be interpreted as political. In addition, it denied members the right to strike or demonstrate. By regulating every aspect of peasants' lives, the state restricted their political power.[12] Heydemann explains:

> [...] the law also brought peasants firmly within the embrace of the state and ensured that neither the collectivization of rural labor nor their integration into popular organization would enhance their political autonomy. The incorporation of peasants into the regime's ruling coalition produced substantive gains in peasants welfare and income, but it followed a thoroughly corporatist design, thus underscoring the regime's authoritarian and state capitalist character.[13]

The law stipulated that all peasants who are recipients of private or public land join a cooperative to get loans or seeds from the state. The new regulations were put in place to prevent the feudal order from dominating again. A vast network of cooperatives was created throughout Syria to make sure that Baathist policies were implemented thoroughly. At the same time, the new system kept peasants under the suffocating grip of the state. Overall, the reform introduced by Nasser's government during the Union weakened the landowner's class. It gave power to peasants but it also built the foundation for an oversized bureaucracy, and opened the door for power centralization and authoritarian politics. By weakening landowners and appropriating their land, Nasser began to gradually integrate rural regions in Syria into the capitalist circuit. The developmentalist project required that peasants support the ruling coalition, which was achieved through the agrarian reform.[14]

Baath: 1963–1970

The ideologues of the Baath party viewed the agrarian reform primarily as a political tool, not simply an economic goal. This means that the primary purpose of the reform is not to generate capital necessary for industrialization but primarily to build rural communities loyal to the regime.[15] Raymond Hinnebusch qualified the Baath rule in 1963–2000 as an "agrarian revolution"[16] due to the centrality of the agrarian reform and the role peasants played in the Baathist coalition. When the Baath party seized power in 1963, it quickly moved to implement an agrarian reform more radical than Nasser's. The neo-Baath, which ruled Syria in 1966–1970, represented the more radical faction of the Baath. It "reduced the ceilings on ownership, accelerated the pace of reform, and ultimately confiscated 22 percent of the cultivated land. Large landowners retained 15 percent of the cultivated area, including much of the best land."[17] The radical and moderate factions of the Baath had substantial ideological differences, including how far the agrarian reform should go. Radicals advocated for state control of vital sectors of the economy. They wanted to undermine the power of merchants and landowners. In April 1964, merchants in several cities organized a strike to oppose the new policies and put an end to their economic and political marginalization. The Baath's response was swift: it quelled the protests and introduced new economic reforms. The state fixed grain prices and began buying peasants' production. These policies weakened the power of merchants in Aleppo and Damascus, many of whom used to stock grains to create shortages, and subsequently inflate prices. Merchants used these strategies to put pressure on the state and undermine its legitimacy but ultimately they failed.[18]

The following section explores the politics of bread in Syria since the 1960s using French philosophers, Gilles Deleuze and Félix Guattari's theory of power. To avoid essentializing state power or reducing it to the coercive force of the military and security apparatuses, the French intellectuals developed a theory of power based on decentralized networks. The history of the agrarian reform in Syria highlights the process through which the Baath built a ubiquitous network that allowed it to control the countryside. Deleuze and Guattari provide a framework to analyze power that is dispersed throughout the social space. Their theory rejects essences to examine instead the ways elements are constructed and ultimately form an assemblage. It allows for a non-conformist reading of history that emphasizes "the becoming" rather than "the being" of objects and subjects. The main question for them is not to discover the inner essence of things but to understand how they connect and operate together. According to Thomas Nail, the questions that Deleuze and Guattari ask:

> [...] are not question of essence, but questions of events. An assemblage does not have an essence because it has no eternally necessary defining features, only contingent and singular features. In other words, if we want to know what something is, we cannot presume that what we see is the final product nor that this product is somehow independent of the network of social and historical processes to which it is connected.[19]

Deleuze and Guattari's philosophy of immanence opens new spaces to reflect upon the Syrian uprising. It avoids ontological questions and decenters the state, both of which are necessary moves to make revolutionary processes visible. Deleuze writes "politics precedes being" and in that sense, theory is always contingent upon the ever-changing geography of power and subjectivities. A theory, or what Deleuze calls a plane of consistency, emerges only as a result of the complex assemblage of heterogeneous elements. The plane of consistency is always shifting and adapting to create a new coherence.

Theory, for Deleuze, should not be simply used to interpret the world but rather to transform it. In what follows, I deploy two particularly productive concepts from the Deleuzian toolbox to examine the politics of bread in Syria. The first one is the "assemblage," which does not exist by itself in the world but rather appears according to specific conditions of possibility. The relations that connect the different parts of an assemblage are meaningful. An assemblage emerges as a result of confrontations with antagonistic forces, as well as the synthesis of various processes that converge toward each other.[20] The second concept is the dyad of "deterritorialization and reterritorialization," which represents two distinct moments of an assemblage. Deterritorialization, according to Deleuze, is the process of change or breaking away from a previous assemblage. It represents the process of liberation from a prior function. The process of reterritorialization refers to the ways in which (deterritorialized) processes form a new assemblage, which is different from the one that preceded it.

The assemblage could be the result of state or capitalist power.[21] The state captures new social spaces by imposing specific relations and logic, which dominate if not challenged. The state builds an assemblage on the ruins of what preceded it. It destroys any processes or networks that do not fit neatly within its assemblage. The state assemblage is totalizing and imposes a closed circuit that gradually absorbs the outside and prevents the inside from escaping it. Deleuze and Guattari write in *A Thousand Plateaux* (1987),

> The State ... makes points resonate together, points that are not necessarily already town-poles but very diverse points of order, geographic, ethnic, linguistic, moral, economic, technological particularities. It makes the town resonate with the countryside. It operates by stratification; in other words, it forms a vertical, hierarchized aggregate that spans the horizontal lines in a dimension of depth. In retaining given elements, it necessarily cuts off their relations with other elements, which become exterior, it inhibits, slows down, or controls those relations; if the State has a circuit of its own, it is an internal circuit dependent primarily upon resonance, it is a zone of recurrence that isolates itself from the remainder of the network, even if in order to do so it must exert even stricter controls over its relations with that remainder.[22]

Deleuze and Guattari reject the idea that the despotic state is simply the result of centralized power. Instead, they propose an impersonal and decentralized understanding of power that is pervasive throughout the social body. The assemblage connects heterogeneous elements to each other to create a seamless relationship between the different parts. For the purpose of this chapter, the assemblage includes the modalities of land distribution, the design of irrigation networks, as well as the construction of dams throughout the Syrian territory. The state assemblage strives to constantly colonize new spaces. Instead of reducing the state to the centralized power of the autocrat, the analysis below shows that the agrarian reform, as an assemblage, is regulated by policies and composed of infrastructures dispersed through the territory. In that regard, state power cannot be reduced to military, the security branches, or the carceral archipelago. It is also, and more importantly, the power of impersonal networks that the state builds by relentlessly colonizing various social spaces.

From a Deleuzian perspective, the Baath agrarian reform should be understood as a process of destruction (or deterritorializing) of the feudal system and imposition of state developmentalism (reterritorializing). The goal of the Baath was to undermine the power of merchants, industrialists, and landowners, and they did so through land reform and by nationalization of the economy's main productive sectors, including the textile and extractive industries.[23] After 1965, the state controlled most strategic industrial sectors, thus undermining the economic power of wealthy classes. Agriculture was evidently necessary to provide the raw material as well as the capital required for various sectors of the industry. It became clear to the Baath that without a robust agricultural sector, state developmentalism was bound to fail. Agriculture was gradually integrated to the structure of the economy, through agro-related industries. In addition, agricultural production was used in local industries such as the textile industry. However, as Italian historian Massimiliano Trentin explains:

> [...] the marginality of the private sector, lack of funds, and mismanagement often led to failures in the heavy industry sector, so that the regime later directed most investments toward agricultural production and to light and transformation industries. Since the state was the main agent for industrialization, the Ba'thists saw planning as the best strategy to rationalize its engagement.[24]

When Assad seized power in 1970, he gradually compromised with the private sector at the expense of the peasants and working classes.[25]

The goal of the Baath was not simply a productivist one—it was primarily about the consolidation of state control over rural areas. While the state owned only a small segment of the land, it controlled agricultural activity through the credit system, the distribution of seeds, or the purchase of grain production. The Baath party built an extensive network, which forced peasants to interact with the state at every level in order to operate. The main purpose was therefore the formation of a populist state, while land reform was one of the tools used to achieve that goal.[26] The land reform allowed Baathists to dismantle the feudal state and erect a new assemblage controlled by the party. It captured deterritorialized peasants and incorporated them into the state assemblage. Peasants escaped one assemblage, namely, the feudal, to be incorporated into the assemblage of the despotic state.

The Baath party built an alliance initially with the poorest segment of peasantry, by furthering their interests through the land reform. The priorities of the party gradually shifted toward supporting the interests of the middle class and landed oligarchy. Describing the regime policies in northern Syria, geographer Myriam Ababsa explains,

> the Ba'thist regimes adopted a pragmatic policy toward the Jazira, which consisted in promoting the emergence of a class of middle-sized shawi landowners who were loyal supporters of the party, while allowing the great feudal landowners to keep the basis of their wealth.[27]

Baath bureaucrats issued an amendment to the land reform that made it possible for landowners who had recently irrigated their land to maintain it. In addition, those who had a close relationship with the party were given some of the most arable land. The purpose of the new legislation was to build a basis of support among the middle-class peasants, who formed a large segment of al-Jazira region in Northeastern Syria.[28] "Their aim was to control a region whose inhabitants were 92 percent rural and 96 percent illiterate, and to create favorable conditions for the implementation of the great Euphrates and Khabur Project."[29] What happened instead is that those who were close to the Baath were able to negotiate good deals while others were excluded.[30]

The Baath expanded the network of the Agricultural Cooperative Bank by creating new branches in many regions including the most secluded ones. The number of branches doubled in 30 years, going from 30 to 62, between 1953 and 1986.[31] In 1964, the regime created the General Peasants Union (GPU), which was led by Baath loyalists, and run as a top-down organization and with no input from members. By the end of the 1960s, the Union had 120,000 members and associations in the vast majority of villages and cities in Syria. The state used the Union to limit peasants' autonomy and keep a close watch on what they were doing. It was used extensively to implement the government's five-year plans, and collect peasants' grain production.[32] The GPU was an effective tool to reterritorialize peasants and expand the reach of the state. While the GPU provided necessary vital resources to the peasants, it also colonized every aspect of their lives. When the land reform was completed in 1970, approximately 1.5 million hectares had been expropriated and redistributed. About one-third went to individuals and one-quarter was used to create peasants' cooperatives.[33]

In 1970, Assad seized power through a coup he labeled the "Corrective Movement" against Neo-Baath's deviations. He threw his political enemies in jail, namely, Salah Jadid and Nureddin al-Atassi, and ended economic policies he considered too far to the left. He stopped the distribution of new land and shifted the Baath's center of gravity from small- to middle-class peasants.[34] In addition, he allowed members or individuals affiliated with the Baath to profit from state institutions. In the end, the regime built a clientist system that rewarded loyalty to Assad. Bureaucrats made deals under the table or used institutional power to profit from a corrupt system. They enriched themselves by leasing state-owned land outside institutional channels, or seizing tribal land.[35] These practices reproduced certain aspects of the feudal system that the agrarian reform was supposed to destroy.

The Baath implemented an agrarian reform in the 1960s to dismantle the feudal system, and the oppressive social relations in the countryside. In less than 20 years, a new cruel arrangement was erected on the ruins of the old one. Hanna Batatu explains that the class structure the Baath created in rural areas is quite complex. It is composed of a range of different groups including wealthy farmers who owned most of the land and communities who earned meager incomes from agriculture. He notes that rural groups:

> [...] consist essentially of three elements. One is composed of the remnants of the old Beys or their descendants. They live in the cities

but thrive in some villages because they are, in the words of the local peasants, mad'umin—supported—from above. Another constituent of the class of richer farmers is rural and has a partly plutocratic and partly official character in the sense that its position is based partly on money and partly on its links with the government or the Ba'th party.[36]

The third group is composed of the investors (*mustathmir*s), who have capital to invest in agriculture and can be of urban or rural origins.[37]

The new social constellation in the countryside was the result of various contradictions within the Baath party. There was an ongoing debate between bureaucrats trained in the Soviet Union, who wanted to maintain state-owned farms, and those educated in the West, who pushed for privatization.[38] The conservative faction of the Baath advocated for privatization of state cooperatives but Assad refused.[39] The regime opposed extensive privatization to maintain economic leverage over Damascus and Aleppo bourgeoisie. Assad feared that extensive privatization would produce a powerful capitalist class that would eventually challenge, or worse, threaten his rule. He preferred to give the Syrian bourgeoisie gradual access to key sectors of the economy and as such prevent them from contesting his power.[40]

In the early 1980s, the security and military apparatuses committed massacres in Hama and Aleppo in order to crush the opposition to the regime, which was in part composed of a debased urban bourgeoisie. By the mid-1980s, Assad had neutralized his most threatening political enemies and believed the moment was opportune for an economic détente with the Syrian bourgeoisie. The regime passed Decree No. 10 in February 1986 to create a friendly environment for foreign investment in agriculture.[41] In 1987, Syria experienced an acute economic crisis. The Baath conservative branch seized the opportunity to push for extensive privatization. In the early 1990s, with the fall of the Soviet Union, the conservative position of the Western-educated bureaucrats finally became dominant, and the regime began a period of economic liberalization that intensified when Bashar al-Assad came to power in 2000. Privatization was convenient because the government did not have the financial means to fund additional state farms or even to maintain the already existing ones because they were draining state resources. Instead, it leased some of the land to entrepreneurs and reserved its resources for building large hydrological projects.[42]

A transition from state to capitalist assemblage occurs, according to Deleuze, when the process of capturing new spaces becomes less consequential for state power. In addition, Deleuze and Guattari write, "[c]apitalism forms when the flow of unqualified wealth encounters the flow of unqualified labor and conjugates with it."[43] While state assemblage assigns rigid roles to individuals and prevents their movement, capitalism "liberates" and allows them to move freely.

THE POLITICAL ECONOMY OF BREAD IN SYRIA

This section analyzes the political economy of bread in Syria since the Baath party rise to power in 1963. It provides a brief historical account of the political, economic, and environmental cost of providing low-priced bread to pacify the population. Agriculture was the main economic activity in Syria until the mid-1970s. By 2010, it still represented 15 percent of the GDP and 800,000 worked in this sector of the economy, representing 17 percent of the labor force.[44] Prior to the uprising, the livelihood of 80 percent of the rural population, representing 8 million Syrians, depended on agriculture.[45] When the Baath party took power in 1963, seven years before Hafez al-Assad's *coup d'état*, one of its priorities was to alter the food economy in Syria. Wheat and cotton became the two most important crops for the government. Cotton was a cash crop that brought much-needed hard currency, at least up until the industrial extraction of oil in 1974. Wheat had also a strategic importance but for different reasons. The regime's goal was to maintain basic commodities at a low cost to prevent urban rebellions. Since bread provides 40 percent of caloric consumption of the average family,[46] the regime turned the production of wheat into a national priority. It gradually created self-sufficiency in wheat production and reduced reliance on imports. It allocated important resources and changed the structure of agriculture to achieve that aim.[47] It enacted changes at different levels in the chain of production to reduce the cost of bread for the consumer. By the mid-1990s, the regime reached its goal of producing enough wheat for domestic consumption but the economic and environmental costs were high. Starting in the 1990s, the Syrian state began exporting some of the wheat surplus.[48]

The state utilized multiple strategies to encourage farmers to grow wheat, including the expansion of irrigated areas, subsidizing seeds and fertilizers, and buying peasants' production at a fixed premium cost. It focused its attention on four northern provinces, namely, al-Hasakah, Deir az-Zor, ar-Raqqa, and Aleppo, which produced three-quarters of the wheat. Despite the Baath party's decision to increase the production of wheat to reach self-sufficiency, the total output was fluctuating enormously up until 1989. During the period of instable production, it reached a low of 0.56 million tonnes in 1966, and a high of 2.24 million tonnes in 1980.[49] Starting in 1989, the output increased steadily to reach self-sufficiency in 1995 with a production in excess of 4 million tonnes.[50] Over the following 20 years, the production was mostly stable, up until the Syrian uprising in 2011.

To increase wheat production, the state expanded the irrigated areas and encouraged farmers to grow wheat instead of other crops. The Syrian state embarked on an ambitious developmentalist program and made the construction of dams one of its cornerstones. In 1963, when the Baath party seized power, there were no dams in Syria. By 2001, their number rose to 160, and they mostly provided irrigation water for agriculture, while some supplied electricity and water to households. Approximately one-quarter of the budget was spent on agriculture, a large segment of which was allocated to building large hydrological projects.[51] The construction of the Euphrates Dam, one of the largest projects, began in 1968 and was completed in 1973. These hydrological infrastructures are not simply economic but primarily political. Geographer Jessica Barnes explains,

> [w]hile portrayed as a technical project, dam construction also has political dimensions. The ability to rearrange the natural and social environment through dam construction demonstrates the strength of the modern state as a techno-economic power. The large Euphrates Dam exemplifies this political motive, designed as a showpiece of Ba'thism to demonstrate the engineering prowess of the Ba'thist state. In addition, the concentration of water resources behind a dam creates a site that can be politically controlled. The police guard tower located at each major irrigation infrastructure in Syria is a visual manifestation of this political control.[52]

As Barnes suggests, the power of the state should not be reduced to its military or police forces. It is essential to understand the networks of irrigation as well as the construction of dams as part of the technologies of power that the state deployed effectively in Syria since the mid-1960s.

In 1998, wheat was cultivated on over 34 percent of the total agricultural land. Thirty percent of the land was irrigated and produced approximately 60 percent of the wheat, while the remaining 70 percent depended on precipitation. In 1990, most irrigated land, around 64 percent, was used to grow wheat and cotton. This percentage increased to 80 percent in 2000.[53] In addition, the state encouraged farmers growing cotton on irrigated land, to turn their attention to wheat production. During the first phase of liberalization in the late 1980s, all restrictions on well construction were removed, doubling their number in a few years. In 2000, when Bashar al-Assad launched the second phase of liberalization, the same phenomenon occurred again. The regime allowed the number of wells to grow from approximately 130,000 in 2000 to 230,000 in 2010. Most of these wells were located in the north, the region of most wheat production.[54] These wells had devastating ecological effects but were not enough to hamper the impact of the last wave of drought, which lasted several years, in 2006–2009. The impact of the drought was drastic and led to the loss of 800,000 jobs, which in turn triggered massive internal displacement. Many of the disfranchised peasants ended up living in informal housing and slums in the suburbs of Damascus and Aleppo. This massive displacement of population was combined with economic liberalization that Assad initiated in the late 1980s, and which his son accelerated in the early 2000s. The drought destroyed innumerable jobs, while economic liberalization increased the price of basic commodities. In addition, it explains the process of deterritorialization of peasants, who had been previously captured by the state assemblage, starting in the 1960s, but who were then uprooted by the drought and economic liberalization. The capitalist logic of the late 1980s and early 2000s freed many peasants and turned them into mobile cheap labor. Many ended up living in informal housing on the peripheries of large cities. The peasants, who initially benefited from land reform and were considered the loyal base of the regime, were gradually marginalized to the point of becoming major actors in the geography of the revolts since 2011.

FROM WHEAT PRODUCTION TO BREAD SUPPLY

To organize the production of bread, the Syrian state created the General Company for Mills (GCM) in 1975, while the General Company for Baking (GCB) was established a decade later, in 1986.[55] The creation of the GCB coincides with the first wave of liberalization of the economy. While the regime began reversing land reform, and allowing private capital to enter the countryside, it also wanted to maintain its control of bread prices through the GCB. The creation of the GCB in the mid-1980s shows that the state had shifted its focus from wheat production to bread supply at a subsidized price. Instead of producing the wheat itself, the Syrian state began subsidizing peasants' production extensively. The GCM bought an average of 2.5 million tonnes of wheat per year from farmers and kept a strategic stock of 3 million tonnes[56] in its silos for food security. The price of wheat was determined by the government and has always been much higher than the price on the international market. The GCM, which could process up to 1.8 million tonnes of flour per year, began contracting private mills to process the remaining quantity, which was needed to provide bread to a growing population. In 2000, for example, it contracted 13 private mills, mostly located in Aleppo, Hama, and Homs to produce 385,000 tonnes of flour.[57] The GCB sold flour at a subsidized price to public and private bakeries, which were required to provide bread to the consumer at a fixed price. The price of bread was fixed by the government, regardless of the price of wheat. In 1999, the GCM spent $256.03 million to subsidize the wheat and provide it at a low price to bakeries.[58] These subsidies represented 3.8 percent of the GDP that same year.[59]

The purpose of the agricultural infrastructure is to entrap peasants and reduce their autonomy to a maximum. The extensive construction of dams and other projects linked to agriculture, namely, agricultural chemicals, networks of mills, and bakeries, should be understood as the consolidation of the Baath's infrastructural power. This form of power is much more effective than coercive force because it restricts the realm of the possible for peasants while also dissuading them from rebelling. The rural population is by definition difficult to control since it is dispersed throughout the territory and is far from urban centers where the might of the state is concentrated. Access to agricultural resources, such as irrigation water or fuel for wells, required loyalty to the regime, whether it was real or simulated.[60]

THE POLITICS OF BREAD SINCE 2011

The Syrian regime faced a major bread crisis when the Uprising began. A number of factors should be evoked here to explain this critical and multilayered crisis. There are at least three important factors that help understand the bread crisis. The first challenge the regime faced was the loss of Northern Syria where 70–80 percent of the wheat is produced. Second, the government diverted most of its resources to the war effort and as such was unable to subsidize the bread economy as it used to in the past. Finally, Syria faced an ecological crisis due to overexploitation of the land and water resources. The impact of this crisis on agriculture was drastic.

In 2014, the land available for agriculture dropped from 1.7 to 1.2 million hectares due to the ongoing war. That same year, Syria's wheat production was one of the worst in recent years dropping below the 3 million tonnes threshold, which had happened only twice: between 1995, when self-sufficiency was achieved, and 2011, when the revolt began.[61] Since 2011, the Syrian government has lost most of Northern Syria, which represents the cereal belt of the country. This region is Syria's food basket and used to produce around 80 percent of the wheat. In 2012, it was forced to import on average 100,000 tonnes of wheat every month.[62] The combined impact of these developments led to a steep increase in bread prices in regime-controlled areas.[63] Since 2011, the Assad regime abandoned the idea of wheat being a strategic commodity. The policy was first introduced in 1975 and remained of central importance until the eruption of the revolt. From that time, the Syrian government stopped allocating substantial resources to maintain control of the cereal basket in the North. In 2015, it announced that it would focus its efforts on maintaining control over "Useful Syria,"[64] a territory the regime deems vital and which includes Aleppo and Damascus, as well as the coastal areas located between the two cities. Since Northern Syria is outside the perimeter of Useful Syria, evidently the regime did not consider its agricultural production was worth fighting for. In 1975, the regime declared wheat as a strategic commodity, and food security as a priority. It developed policies and built massive irrigation networks, and many dams to achieve that goal. When the population rebelled in 2011, the regime turned wheat from a strategic commodity into a weapon of mass destruction. Not only did it cease to provide it to a large segment of the population, in addition, it burnt wheat fields on a large scale.[65]

The regime's war efforts have prevented it from utilizing a large share of the budget to subsidize bread. As mentioned above, for several decades, the Syrian state used to pay a premium price to farmers and buy their wheat production. The state imported wheat only when the strategic stock fell under a certain threshold value. State protectionism was maintained for several decades and it gave Syrian farmers the incentive to produce wheat without the fear of losing money, since the production was always bought by the state with a margin of profit varying between 20–30 percent. This policy maintained the price of wheat well above the international price. In the end, what mattered for the Syrian government was to provide bread at a low price.

In addition to these economic factors, the regime was also facing a major ecological crisis. In 2006–2010, there were several droughts, which amplified the crisis of wheat production. Gianluca Serra, who spent a decade working with the Food and Agriculture Organization of the United Nations (FAO) in Syria, explains that the regime used the droughts as a pretext to avoid taking responsibility for the decline in wheat production. In reality, the droughts were due in large part to the state's reckless management of the land and water as well as the ecological crises that ensued. He explains that the crisis:

> ... began in 1958 when the former Bedouin commons were opened up to unrestricted grazing. That led to a wider ecological, hydrological and agricultural collapse, and then to a "rural intifada" of farmers and nomads no longer able to support themselves.[66]

The Syrian regime conveniently used the argument of natural drought to deny the over-exploitation of the land and refute taking responsibility for the current crisis. A number of studies have shown that over-exploitation, the construction of wells, and other policies encouraged by the state, have led to the desertification of large territories.[67]

MANBIJ: A CASE STUDY

In July 2012, revolutionary forces liberated Manbij peacefully, a city of 200,000 inhabitants in Northern Syria. For almost a year, neighborhood groups had organized small and large protests and creative peaceful actions until they managed to expel security and police forces from their city. After liberation, they formed a revolutionary council and began working relentlessly to make their city liveable, despite the ongoing violence they were facing. The city and its inhabitants reinvented every institution and came up with creative ways to solve everyday problems. The revolutionary council and active groups in the city began a process of de-Baathification by deploying a combination of traditional knowledge and decolonial practices. By governing the city, they also faced a myriad of problems in a context of acute poverty and weekly airstrikes whose aim was to prevent normal life after liberation. The politics of bread in Manbij is important because it highlights the challenges facing revolutionary groups as they attempt to create a self-sufficient economy. The production and distribution of bread outside the regime's networks constituted a direct threat to its survival. For almost half a century, the regime imposed a simple equation in Syria, according to which cheap bread would be available as long as the population refrains from participating in the political process or criticizing the Assad regime. It built a complex infrastructure that includes dams, irrigation networks, seeds, fertilizers, as well as mills and bakeries to control every aspect of the bread economy. It made it virtually impossible for peasants to escape its infrastructural power, while consumers were provided with cheap bread. After 2011, cities such as Manbij began to challenge the regime's narrative by building autonomous circuits of bread. The production of bread in the liberated areas was threatening the regime's infrastructural power and as such were often targeted by its jets. Building a bread economy outside the government networks was simply not permitted. The creation of an independent circuit of bread is a challenging task. Not only does it require the deterritorialization of processes that have been in place for more than half a century, there is also need for a nomadic assemblage that reterritorializes the new processes. Revolutionaries have to dismantle the state assemblage and create a more democratic and egalitarian one. In the section below, I explore the micropolitics of bread and the ways it was operationalized in the city. The politics of bread sheds light on the strategies adopted by the population to delink from the regime's ubiquitous infrastructural power. In addition, it shows how the regime responded to prevent the emergence of a nomadic assemblage outside the purview of the state. Thomas Nail explains that a nomadic assemblage "constructs a participatory arrangement in which all the elements of the assemblage enter into an open feedback loop in which the condition, elements, and agents all participate equally in the process of transformation."[68] In what follows, I present some of the strategies and counter-strategies deployed in the 18-month period (July 2012–January 2014), during which the revolutionary council controlled the city.

Bread production and distribution

Manbij has one of the largest flourmills in Northern Syria, making the city vital and strategic for the entire region. The mills can process up to 450 tonnes of wheat a day, a quantity sufficient for 1 million inhabitants. After the liberation of the city, the regime kept providing Manbij with wheat to maintain the state assemblage in place. In addition, the director of the mills, and approximately 100 employees were kept on the regime's payroll. The revolutionary council was initially unable to provide wheat at a low cost, or pay salaries, and as a result was forced to accept the regime's indirect presence in the city and the dangerous consequences of such a policy. While Manbij had dozens of brigades fighting the regime, it did not have the resources to completely dismantle the state assemblage present in the city. When the director of the mills threatened to leave due to repeated disputes with various powerful actors in the city, the revolutionary council created a team of volunteers to shadow the mills' technicians and engineers and gain the necessary skills to operate the mills independently and avoid a possible starvation of the population in the event the director or the employees decided to leave.

Targeting bakeries

In December 2012, the regime began an airstrike campaign in several cities targeting people waiting at bread lines in front of bakeries and gas stations. The violence terrorized the population and undermined post-Assad alternatives in liberated areas.[69] To counter the regime's indiscriminate killing machine, the revolutionary council in Manbij, after deliberations with various actors in the city, decided to distribute bread in different neighborhoods to avoid gatherings in front of bakeries. It hired a large number of young men looking for jobs and assigned them to different neighborhoods to distribute bread. To prevent the selling of bread on the black market at exorbitant prices, the council gathered extensive data about the number of families and their needs in every neighborhood and rationed bread accordingly. This process allowed the council to decentralize the distribution of bread and hence end long waiting times in front of bakeries as well as prevent the regime from attacking crowds at these locations. The bread census provided vital spatial data about the city and its inhabitants. The new process prevented individuals from buying large amounts of bread and selling it at a higher price on the black market. One of the setbacks, however, is that new refugees were not accounted for in the census and, thus, could not buy subsidized bread; as a result, they were forced to pay twice or three times the price of bread on the black market. These processes demonstrate the difficulty of erecting a nomadic assemblage in the liberated regions.

The geography of bread

Mills and bakeries were vital institutions under Assad's rule but they became more so in the liberated areas since 2011, as bread grew into a crucial staple for most Syrians, who often relied on it for their survival. With the liberation of Manbij, the revolutionary council made the provision and distribution of bread, and the protection of the city from the regime's ground attacks, a main priority. In reality, bread and freedom are inseparable: the liberation of a city is meaningless in the eyes of its inhabitants, if the living conditions worsen as a result. The council was fully aware that success or failure depended on whether it could provide bread at the same price as in regions controlled by the regime. Likewise, the regime understood the revolution would be meaningless if it is unable to provide bread to the populations living in the liberated areas. The revolutionary council created a special committee to examine the various scenarios and propose strategies to make bread available at a low cost. It was able to solve the problem of bread sold on the black market but was still facing other major problems. After the liberation of the city, large numbers of military brigades were formed to fight the regime. In some cases, powerful families and clans created their own groups to defend their interests and maintain influence in the city and surrounding areas. These brigades were known as "the bread brigades" as they did not fight the regime but had the same privileges as the ones who did. FSA groups could not wait in line for long hours to get bread because they needed to return to the front promptly. The population in Manbij was critical of the bread brigades but was unable to stop them.

Another issue arose quickly after the liberation of the city. The revolutionary council wanted to prevent powerful military groups present in the city from controlling the mills and monopolizing the distribution of bread. The mills were difficult to guard as they were located on the outskirts of the city, making them vulnerable to attacks and as such, an easy target for belligerent military groups. For example, when Ahrar al-Sham, a powerful jihadist group, forcefully took control of the mills under the pretext that the management was corrupt and lacked financial transparency. The council then organized a successful campaign to get them out of the mills. By providing bread at a low cost, the leader of Ahrar al-Sham believed his group would gain the population's loyalty. His plan backfired, however, as the entire city opposed military involvement in civilian affairs and did not approve of the forceful way the group had taken the mills. The council and several powerful groups in the city put all their differences aside and organized several protests until Ahrar al-Sham were forced to leave the mills.

From the local to the regional

To end the city's dependency on the regime, the revolutionary council began building an alternative circuit of wheat in the liberated regions by creating a geography of solidarity. This new geography required the liberated cities to share the benefits and burdens of dividing up the wheat equally among the different regions. The new circuit of wheat did not always work smoothly because revolutionary councils in the liberated region had to maintain a subtle balance between specific local demands and an elusive regional strategy. For example, Raqqa's local council refused to lend Manbij its expensive equipment to fix a power cut, despite the good relationship between the two cities. The former feared that a corrupt FSA group would steal the equipment at a checkpoint between the two cities. Manbij's revolutionary council responded by threatening to cut the water supply and stop providing bread to Raqqa's western countryside, thus putting enough pressure on the neighboring city to finally lend its equipment. The scarcity of resources and the presence of multiple military groups with divergent agendas made cooperation between various revolutionary councils a main challenge. This incident, and others, shows that democratic governance at a local level is not easily transposable to a larger regional scale.

Bread as a weapon of war

While the revolutionary processes were producing a nomadic assemblage in the city, the regime was brutally destroying it. The regime used bread to strangle the revolutionary process in Manbij. When it was not able to control the economy of bread in the city anymore, it began targeting it violently. For example, the regime began bombing bread lines in front of bakeries in summer 2012.[70] It besieged areas controlled by the opposition and starved the communities inhabiting them. In addition, the regime bought wheat from areas controlled by the opposition by proposing a higher price. It was granted a line of credit by the Iranian regime to do so. According to the most recent estimates, more than 50 percent of the areas that produce wheat were under the control of the opposition in 2015. Some farmers sell their crops to a middleman who then smuggles them to a neighboring country such as Iraq or Turkey.[71] The regime targeted the Rashediah food warehouse and regularly burned wheat production in areas controlled by the opposition. These processes, in addition to the ongoing war, left more than 50 percent of the Syrian population in need of food assistance in 2015, according to FAO. This number was only at 5 percent before the revolution.[72]

CONCLUSION

The chapter examines the history of the agrarian reform and its implications on the rural population. Before the revolts, the regime built a politics of bread that would prevent dissent and contain rebellion. The chapter illustrates the power of infrastructural control, which refers to the infrastructure used for the production of bread. It includes dams, irrigation networks, seeds, fertilizers, bakeries, and mills, and which together form the state assemblage. This form of decentralized power captures spaces that were previously outside its preview and turns them into networks of control. The regime created a state assemblage in the countryside, which produced dependency among the peasantry. As a result, all aspects of agriculture depended on the state, while peasants could not survive without the assemblage. This form of power is more effective than military force as it requires little manpower to maintain. Land distribution destroyed the feudal order but it simultaneously created a peasantry that is highly dependent on state resources (dams, seeds, fertilizers, irrigation networks, credit, etc...). The state logic progressively incorporated the market logic. It gradually substituted the state assemblage with a capitalist assemblage, especially when Bashar al-Assad seized power in 2000. The Baath began reversing the land reform and large segments of the peasantry was forced to leave the countryside and join the disposable industrial reserve army in the large cities.

The state and capitalist assemblages became ineffective when the 2011 revolt erupted. The regime tried to contain the rebellion but as its scope broadened, Assad began deploying a lethal strategy aimed at eliminating all forms of opposition. Wheat, which was used to pacify the population in the past several decades, quickly became a weapon of mass destruction—and one in which bread played a central role. Finally, revolutionaries in Manbij developed strategies to delink the economy of bread from the state assemblage. Despite the revolutionary council's failure to build a sustainable economy, the practices that emerged around the production, circulation, and consumption of bread provide valuable lessons for the future.

5
Participatory Democracy and Micropolitics in Manbij: An Unthinkable Revolution[1]

You can jail a revolutionary, but you cannot jail a revolution
                                    Syrian Rebel Youth, Homs, July 24, 2013

Manbij, a city in Northern Syria, hosted a compelling example of successful grassroots governance during the two-year period between the Syrian regime's withdrawal from the city in 2012 and the Islamic State's takeover in 2014. The city established an innovative local political system during this interregnum. The new local government faced significant challenges and made many mistakes, which I discuss in detail. Those mistakes, however, were not the undoing of Manbij's revolutionaries. Instead, as in many other places in Syria, external forces derailed their efforts, buoyed by a Western narrative that seemed unable to even conceive of the kind of peaceful revolution under construction in Manbij. Still, Manbij's experience holds important lessons—and could yet be the foundation for more participatory governance in Syria over the long term.

Twenty miles south of the Turkish border, on a plateau to the west of the Euphrates, lies the Syrian city of Manbij.[2] With a population of 200,000, it is located halfway between the cities of Aleppo and Raqqa, the latter of which was the de facto capital of the Islamic State. Manbij is a millennia-old city with a history, like so many urban places in Syria, of religious and ethnic diversity; it has passed through the hands of many empires, ranging from the Assyrians to the Ottomans. Today, the population is predominantly Sunni Arab but also includes several other religious and ethnic groups, including a sizable Kurdish minority.[3]

This place in the province of Aleppo was the setting for a remarkable 18-month period of participatory democracy, from July 2012 to January 2014, during which an entity called the Revolutionary Council (RC) controlled the city. In 2013, I was able to spend some three months in

Manbij over the course of two visits, and witness first-hand its experiment in self-determined governance. My research opened a window onto a completely new perspective on Syrian revolutionaries (most refuse to describe themselves simply as members of the opposition)—one that completely bucked the simplistic mainstream narratives of Western media.[4] Here, during an interlude of relative stability when neither the Syrian government nor foreign-backed jihadist groups had taken over the city, Syrians were deciding how to run their city for themselves—and they were excelling at it.

More than five hours from Damascus by road in peacetime, Manbij was relatively early in throwing off the yoke of Syrian president Bashar al-Assad's regime, who local protesters managed to drive out in 2012. Despite the violence of the regime and military groups such as the Islamic State (which ultimately expelled the RC), and the intervention of a large number of regional and international players in the conflict, the 18 months of locally guided government in Manbij were positive proof that Syrian society was able to produce an original democratic culture and creative alternative governance institutions that were vital to solving everyday problems.

Syrian revolutionaries like those in Manbij did not speak the dominant academic Western language about social processes and revolutions—either because they could not or were unwilling to. As a result, many journalists and academics have effectively denied them any form of agency. Many Western descriptions present them as mindless fighters who are easily manipulated by the different regional and global powers. This dominant narrative is a Eurocentric one that presents the revolutionaries as voiceless victims blown hither and thither by the winds of larger political processes.

The absence of discourses about grassroots processes in the revolution, however, does not mean that those processes are not there. Rather, it is a result of deficiencies in analysis. Too often, the metanarrative of "secular tyrant versus bumbling Muslim insurgents" has blinded scholars and pundits to the possibility of the uprising's piecemeal but significant achievements. A successful revolution became "unthinkable"[5] to Western experts long before external forces had actually broken its back.

This report is an attempt to stitch together a narrative that more adequately represents the grassroots efforts at governance that emerged in 2012–2014: they may have been fragmented, but they also formed an effective and constantly evolving whole. Revolutionaries actively experimented with new conceptual tools and with building alternative institutions on the ruins of the old ones. The revolutionary actions that might appear chaotic from afar acquire new meaning when scrutinized from the perspective of the population that they directly affected.

THE REVOLUTIONARY COUNCIL

In the first half of 2012, a year after protests had swept through Syria and many months after the protests had turned into war, Manbij was still under control of the regime of Bashar al-Assad. Then, in May 2012, the government and allied militias massacred more than 100 civilians in the town of Houla in northwestern Homs province. The incident was not particularly close to Manbij, but the slaughter, which counted dozens of women and children among its victims, galvanized anti-regime protesters around the country. In response, the residents of Manbij organized protests. Marginal at first, they evolved into a general strike and open defiance of the police and security forces. The activists behind these protests came from different political backgrounds. Some of them were secular and others religious, but all utilized non-violent strategies. Activists also created an FSA group—the FSA had been established nearly a year before—but its main function was to fight the regime in Aleppo and other cities.[6] After the strike, which led to the shutdown of the entire city, the protests became larger and overwhelmed the security and police forces, who ultimately fled the city in July 2012.

The FSA gave pursuit, but that was not the main factor in the security forces' departure. News reports said that rebels had "captured" Manbij, but the reality is that there were only a few dozen fighters in the city. The peaceful movement was the real catalyst.[7]

The RC had been founded in secret in April 2012, three months before the revolutionaries gained control of Manbij. During the initial period, it mostly organized protests to expel the police and security forces, and coordinated with local and revolutionary councils in other cities. The members of the RC had diverse backgrounds, but the majority had a university education and came from middle-class families. Several had been Syrian Baath party members before March 2011. Members included lawyers, engineers, doctors, and teachers, in addition to a few Muslim scholars. The RC marginalized the religious establishment because of the latter's collaboration with the regime and its refusal to take a clear stance in support of the revolution. While the person in charge of organizing the protests was the focal point of the group, others, such as the liaison between the RC and the FSA, played an important role as well. The FSA—which by now counted tens of thousands of fighters and controlled major swathes of the Syrian countryside—still had no significant presence in Manbij due to the vitality and momentum of the peaceful activists, who prevented the militarization of their activities to avoid the destruction of their city.

One of the achievements of the RC before liberation was the establishment of a network of more than 50 popular committees whose role was to protect the city and fill the vacuum on the day the security and the police would be forced to leave the city. So when the regime evacuated Manbij, the revolutionaries took over, and with the help of neighborhood committees, they prevented looting. The revolutionaries visited the different parts of the city to make sure that there were security checkpoints everywhere. The popular committees protected the entire city for three days, and not a single incident of violence or theft came to light. Revolutionaries had learned about such local organizing processes from Egyptian protesters who created their own popular committees in Alexandria, Cairo, and other cities to protect their neighborhoods from the regime's thugs and other criminals during the first stage of the 2011 uprising in Egypt.[8] The RC's tremendous success in organizing protests, forcing the police and security to leave the city, filling the political and administrative vacuum, and protecting residents during the transitional period made it very popular. Despite certain missteps, such as the marginalization of important activists and the inability to raise enough funds, the RC was able to govern for several months without much opposition because its members had symbolic capital and revolutionary legitimacy.

While the RC was able to address many urgent issues with creativity and openness, several groups challenged it because they disagreed with its politics, creating a real fracture in the city. After liberation, there were several centers of influence in Manbij. The RC had the most legitimacy because of its organizing efforts prior to liberation and the risks its members took during that period. It was composed of the representatives of the different neighborhood committees, who were active members of the community prior to the revolution and had a vast network of contacts.

However, there was another center of power, revolving around Mohammed al-Bishir, an ambitious architect who felt marginalized. He did not possess much legitimacy in the streets of Manbij, but was able to build a coalition with the individuals and families that were sidelined—intentionally or accidentally—by the RC. His coalition comprised more than 20 new groups and organizations that formed in 2012 and 2013. Some of them were active and had many members while others were ghost organizations created for the sole purpose of increasing the number of groups on Bishir's roster. Many were professional organizations: journalists, lawyers, doctors, teachers, religious clerics, and engineers. Others represented women, the youth, and the media. The coalition elected Bishir to the presidency of a new local council in December 2012, in an attempt to undermine the popularity of the RC, but they failed to do so.

Besides these two major nodes of power, there were several other groupings with more or less political weight that had an influence on events after the liberation. The Islamic religious institutions were one such grouping, but they remained marginal in 2012 due to their unwillingness to take a clear stance against the regime when the city was still controlled by the regime. Some of the young Muslim clerics were active in the protests but the religious establishment, which feared a backlash from the regime, did not support them.

Family affiliations made for another axis of power in Manbij, especially the clans. Every tribe in the Arab world is composed of several clans with uneven access to resources and power; about 15 percent of Syrians belong to a tribe. Certain clans within the same tribe sometimes had competing interests. Manbij's clans[9] were split in their allegiances, and as such did not have much leverage on their members. Despite most clans' marginal role in the protests, some of them made more substantive contributions. For example, the al-Bou Banna clan had taken a clear position against the regime and many of their members were actively participating in the demonstrations. Most clans' sheikhs, however, were close to the regime and had left Manbij early on, fearing reprisal from the population.

The city's notables formed another cohort of consequence in the city. Local dignitaries who belonged to powerful and respected Manbij families had been used by the regime before 2012 as intermediaries to control the residents. They played an important role during the protests and after liberation, as they were frequently asked to mediate between the security and the protesters. After liberation, the RC consulted them whenever it was time to make an important decision.

In addition to operating within a multipolar socio-political scene, the RC faced three main challenges after the liberation of the city. First, it did not have any financial resources to run vital institutions. Second, it did not have the necessary expertise to operate the mills or to administer the various branches of the council. Finally, the RC was unable to prevent the creation of numerous military formations. Powerful families and wealthy individuals in Manbij could form their own FSA brigade. On the day of liberation, there were two FSA groups, then, just three weeks later, their number had increased to more than 30, and by 2013, there were 70. Also, the RC was losing legitimacy because of its inability to effectively protect the population from the harassment of certain FSA groups.

The presence of various military groups in the city caused old feuds to resurface, which prompted certain families and clans to establish their own military formations. While most of these formations were fighting Assad's regime, a sizable number were coercing the locals into paying tributes and taxes. One such group was led by a local thug, who called himself "the Prince" (using the English word). He used his charisma to recruit fighters and later began looting other cities and kidnapping residents for large ransoms.[10] This enabled him to amass the necessary funds to build a large FSA faction that, at its peak, had several hundred fighters.[11] His group was affiliated with the Farouq Brigades, a large grouping of rebels within the FSA,[12] but he avoided a confrontation with the regime to focus his attention on illegal activities and increase the size of his group. Since the Prince did not want to clash with the inhabitants of Manbij, his native city, his kidnapping operation focused on the nearby city of Raqqa. The larger and more powerful FSA brigades disapproved of his actions but were too busy fighting the regime at the front to send fighters to Manbij to arrest or expel him. Ahrar al-Sham, one of the largest and most powerful jihadist groups, finally decided to confront the Prince and use him as an excuse to access Manbij and establish a headquarters there; they entered the city in February 2013. This strategy of coming to the rescue of the inhabitants was not unique to Manbij. Many groups used it to justify their invasion of a city and the subsequent takeover of buildings to establish their headquarters. The al-Nusra Front (which renamed itself Hayat Tahrir al-Sham in January 2017) used the same strategy to establish a presence in Manbij.

The RC was further marginalized when jihadist groups such as Ahrar al-Sham entered the city. While the RC tried to be democratic and inclusive by creating the Council of the Trustees (which we will discuss more below), the revolutionaries were unable to mobilize a large segment of the population. Their isolation allowed opposing forces to mount an effective opposition to the council.

As we have seen, the RC was very powerful immediately after the liberation of Manbij, because of the prior struggles of its members. The revolutionaries built credibility and gained symbolic capital before they took control of the city. Once they had it, they used the legitimacy they had accumulated during the previous period to build their strength. Still, the challenges we have listed stymied them and evolved into even more complicated problems: a lack of funding; a lack of expertise to run institutions; an inability to provide the necessary resources to the internally displaced living in the city; the failure to prevent military groups' intimidation of the residents; an inability to stop the threat and violence of the regime; their unwillingness to create an inclusive space for individuals and groups critical of the revolution; the social pressure of resolving everyday problems quickly (which in turn was difficult because of their other challenges); and the inability to build a solid front against the jihadist groups. The Islamic State capitalized on each one of these weaknesses to capture the institutions one by one and later, the entire city.

COUNCIL OF THE TRUSTEES OF THE REVOLUTION

To regain its legitimacy after numerous setbacks, the RC decided to create a Council of the Trustees of the Revolution (CTR), a legislative formation of 600 residents who had a reputation as revolutionaries. While women were excluded, the CTR did make efforts to be more inclusive than the RC had been thus far. Many new groups and individuals were invited to join. The council had weekly meetings to discuss issues proposed by the members. The CTR made all the important decisions, such as raising the price of bread (the RC controlled an important mill and grain silos in Manbij) and the creation of a police force in the city, while the RC implemented these resolutions.

However, the CTR's attempts at greater inclusion had limits, and the selection of members was controversial: many activists critical of the RC were not invited to join the CTR. While the RC argued that the new council should be open only to revolutionaries (those who had participated in demonstrations, fought the regime forces, or supported the revolution publicly or financially), others wanted to include everyone except for those who had collaborated with the regime or took a clear stance against the revolution.

The CTR's revolutionary–counter-revolutionary binary would come back to bite it. However, in the beginning, tensions gave way and the CTR created a unique space where another kind of politics could be imagined. The meetings facilitated passionate debates about politics, democracy, military strategy, relief work, and various social issues. Any member of the CTR had the opportunity to discuss topics he deemed important. Unlike any other form of political consultation in recent memory in Manbij, the CTR discussed human rights with dignity, in the context of the vernacular culture's notions about respect and piety. The CTR was an experimental space par excellence where grassroots politics, experimental processes, and participatory democracy could thrive. The meetings were organized on a biweekly basis and usually started at seven o'clock in the evening, running until around midnight. They were held at the headquarters of the Revolutionaries of Manbij Battalion on the city's eastern outskirts. Since the regime had information about these meetings through its informants, it sometimes flew jets over them to intimidate participants—and impress upon them the futility of their efforts. Still, the CTR created a public sphere for debate and it was very effective in preventing the FSA and other military groups from controlling the city's institutions.

The CTR gave new momentum to the RC, but it did not last long. The RC made a serious political mistake by excluding a large section of the population. Their minimalist definition of who qualified as "revolutionary" was polarizing. Many in Manbij were insulted by their exclusion from the council, and, consequently, by their implicit categorization as counter-revolutionary. For its part, the RC purported to have made its classification to protect the CTR from the infiltration of counter-revolutionary forces, which were undeniably present in the city. Manbij was the target of weekly airstrikes and many informants provided vital information to the regime. In addition, a section of the population passively supported the regime for a variety of reasons. By imposing such an exclusive definition of "revolutionary," the RC prevented many neutral or passive citizens from joining the CTR. As a result, it was easy for truly counter-revolutionary forces to convince the population that a few families and notables controlled the trustees, while the vast majority of common people were excluded from decision-making.

Groups such as Ahrar al-Sham and the Islamic State began questioning the legitimacy of the CTR and recruiting the residents who felt disenfranchised. Some of the excluded groups gradually built relationships with sympathizers within the CTR, but more importantly, they began organizing against the CTR from the outside. The jihadist groups and their allies used an array of tactics to undermine the legitimacy of the council. They organized lively public debates in mosques about theology and Islam. At night, they held rallies in the city center, closing streets and playing jihadist songs. In some cases, they kidnapped and assassinated members and supporters of the RC. They used intimidation and violence to take over mosques that were under the control of the Sufi[13] establishment and used them as spaces to propagate jihadist ideologies. In several instances, a preacher from the Islamic State would enter the mosque with an explosive belt and a machine gun and tell the imam to leave.

In addition, the Islamic State would ask its supporters to be present to create a threatening environment if the imam refused to obey. In July 2013, the Islamic State kidnapped Sheikh Said Mohammed al-Dibo, the imam of the Grand Mosque, because of his refusal to obey their orders. A large demonstration was organized on the same day and the Islamic State finally released the imam after several factions of the FSA threatened to take action. But the imam was killed early one morning a few weeks later, on his way to the mosque; the Islamic State was considered the most likely assassin. The Islamic State undermined the power of the RC by taking over vital institutions such as the mills and the bakeries. They organized protests against the RC for failing to control the prices of basic commodities or to stabilize the dollar's exchange rate. Finally, they threatened to assassinate the president and members of the RC, and succeeded in creating a climate of fear.

THE REVOLUTIONARY COURT

After the liberation of the city in 2012, the RC started a process of deliberation with various forces in Manbij to establish a court. The Revolutionary Court was created two months later. It was one of the first revolutionary courts in the liberated areas, but the process of setting it up was somewhat contentious. The religious establishment wanted a religious court and a legal code based on Sharia. On the other side were lawyers and revolutionaries who wanted to preserve the Syrian penal code and strip it of any articles created by the Assad regime. Lawyers argued that the Syrian penal code was approved in 1949, almost 20 years before Hafez al-Assad's rise to power, and was thus still useful. In addition, Article 3 of the 1973 Constitution states that "Islamic jurisprudence is a primary source of legislation," which they took to mean that there was no conflict between the penal code and Sharia. Lawyers also argued that only a few articles of the penal code were introduced by either Hafez or Bashar al-Assad's regime, and as such, they could be easily removed. Due to these differences of opinion, it took the Revolutionary Court several months before finding a suitable location to operate, and to agree on a penal code that suited both parties. In the end, a council of lawyers and clerics was formed and instead of employing the Syrian penal code, the RC decided to adopt "the Arab Unified Penal Code," which was drafted by the Arab League in 1996 and was more in step with Sharia than Syria's code was. Practically speaking, however, many lawyers still followed the Syrian penal code since they were familiar with it, and with a few exceptions, Muslim clerics did not disapprove of the rulings, as they had no experience with either.

The Revolutionary Court faced additional challenges—some internal and others external. Internally, the lack of funding prevented the court from having a police force to protect lawyers from retaliation and to enforce the court's rulings. The judges felt vulnerable and unable to inflict harsh rulings—which were viewed as necessary in a time of war, according to some victims. As a result, many of the more serious cases were tried in the Islamic State's religious court. Also, the Revolutionary Court was unable to pay its guards adequately, which made them vulnerable to corruption; some were always seeking extra cash to complement their meager salaries. When families of prisoners began bribing the guards for favors, the power of the Revolutionary Court was undermined.

Externally, the Revolutionary Court's challenges were also severe. For example, some of the larger families and clans in Manbij argued that customary law should apply whenever one of their members was accused of wrongdoing. While clans were weakened after the liberation of Manbij because many sheikhs sided with the regime, some of them still had some influence. In addition, large military factions such as Ahrar al-Sham, the Islamic State, and before them certain FSA groups created their own courts to evade accountability. These courts originally tried fighters who had been accused of wrongdoing. At a later stage, the inhabitants sought them out for other types of cases because they felt that the Revolutionary Court was toothless or too secular.

The multiplicity of courts in Manbij created a conundrum. When a plaintiff or a defendant did not agree with the rulings of the Revolutionary Court, nothing prevented him from requesting a new trial at one of the other courts in the city. An increasing number of inhabitants sought the Islamic State's court because it was much faster in issuing rulings and in implementing them. On several occasions, the Islamic State attacked the Revolutionary Court and captured prisoners. In one instance, the group executed the prisoners publicly to tarnish the reputation of the RC and terrify the population.

Due to these challenges, the RC decided to replace the Revolutionary Court with a Sharia Council in October 2013, which had the support of all the military factions present in the city except for the Islamic State. The RC also created a security brigade to protect the new council and its prison, and to improve security in the city.

But there were now, in reality, three legal systems coexisting in Manbij: the Syrian penal code, Sharia, and customary law. Such a space, where several legal systems coexist, is common when the authority of a central state disappears. The experimentation with this plurality might have been productive in other circumstances, especially since previously marginal groups started having a voice. This was not exactly the case in Manbij, however. Force rather than democratic deliberation imposed certain legal discourses and marginalized others. Still, the existence of several legal systems allowed for creative experimentation, and, in some instances, turned law into an emancipatory rather than a coercive tool.

CONCLUSION

Manbij's democratic experiment was not to last. The rending forces of war proved too strong. During my first visit in June and July 2013, the RC was dominating political life in the city, while the Islamic State (at the time, known as the Islamic State in Iraq and Syria, or ISIS) was marginal. By December 2013, when I went for a second visit, the Islamic State had become the dominant group and controlled most vital institutions, and the RC had almost ceased operating because the lives of its president and members were constantly threatened. As the opposition fought the Islamic State in Northern Syria in 2014, Manbij fell in and out of the extremist group's hands. After being involved in some activities to resist the Islamic State, including a daring general strike, the RC was ultimately expelled and began operating in exile from Azaz, a city more than 100 kilometers to the west, near the Turkish border. In August 2016, the Kurdish Democratic Union Party drove out the Islamic State; the RC seems to have grown much closer to Turkey since the country's military intervention in Syria that same month, and has lost its independence. It has not returned to Manbij.

Manbij's revolutionaries were marginalized by better-funded and extremely well-organized forces. Nevertheless, this is not a sign that Manbij's revolutionary governance was necessarily fated, from the beginning, to be short-lived. The unfortunate demise of the RC's brand of participatory politics was not due to inherent flaws—a failure of its design or implementation—but to external factors.

One of the biggest root causes of the RC's undoing was the revolution's inability to produce a discourse that adequately represented it, in international media and elsewhere. The story has never been told well, either by analysts, pundits, or journalists.

At the international level, the Western and Arab governments have not only intervened militarily to crush the Syrian revolution, they have also produced discourses that besieged the very idea of revolution. The West has advanced its narrative either through complicit silence,[14] well-orchestrated campaigns to tarnish the image of the revolutionaries,[15] or the funding of the most reactionary military factions, such as Ahrar al-Sham and the Islam Army (Jaysh al-Islam).[16] Further, many commentators, journalists, and academics write solely about the military and geopolitical dimensions of the uprisings and ignore the cultures of liberation and participatory politics that Syrians have been developing and enacting since the early days of the revolts. Even when they have good intentions, a large section of the intelligentsia has used antiquated conceptual tools and inadequate theories.[17]

It is important that the violence of the war and the hostile narratives promulgated by outsiders do not silence the stories of local practices such as those in Manbij, which were vital spaces for experimentation and the production of new cultures. The local narratives that emanated from the city stand in stark opposition to abstract global designs and regional strategies that commentators regurgitate when they write about Syria.

Local governance and participatory democracy in Manbij were not the products of ideological certainties but rather the contingent outcomes of grassroots resistance. The efforts were imperfect, but they represented the beginning of a long process of liberation—a process that has been undermined, though not necessarily aborted, by foreign interventions and the Syrian regime's politics of death and destruction. The experimental grassroots politics and micropractices in Manbij and other cities might still represent the beginning of a culture of liberation that could take several decades before reaching fruition. What's more, Manbij's revolutionaries have been aware of their experiment's potential from the beginning, even if they could not have foreseen all the obstacles that would later be thrown in front of them.

As the Syrian civil war finishes its sixth brutal year, a future without Assad is looking ever more unlikely. But there is still wisdom to be gleaned from the efforts of the Manbij revolutionaries, who stumbled not because they made any mistakes in designing their institutions, but because so much of the world turned out to be against them. Throughout the country, revolutionaries' initial experiments with governance—in Manbij, Idlib, the Eastern Ghouta, and elsewhere—will have enduring benefits for the Syrian people, whose skills and aptitude for participatory politics were expanded, however haltingly, in their interregnum of self-determination. If we are to continue to envision the possibility of a more democratic future in Syria, we would be wise to keep the lessons of places like Manbij alive, as well.

Conclusion

The demise of the Syrian revolt is perhaps signaling the end of a revolutionary cycle, while the popular struggles in other parts of the Arab world, namely, Algeria, Iraq, and Sudan are gesturing toward the beginning of a new one. These revolts will take many years, possibly several decades before they reach maturation. In the process, they could potentially alter the socio-political and economic structures in the Arab world. In addition, they are disrupting the geography of knowledge and structures of thinking. As people struggle against dictatorship, common sense is being transformed. A democratic future, which appeared impossible a decade ago, has now become possible, despite its daunting absence. Dictatorship, which until 2011 was the only horizon in Syria, is today undergoing a structural crisis. One of the repulsive Syrian slogans was "Hafez al-Assad, our leader until eternity, and beyond eternity." Struggles against dictatorship, despite their pitfalls, have shown to Syrians that another outcome is possible. In a way, the Syrian uprising announced the demise of Assadist eternity, despite the shortcomings of the revolution.

The five chapters explore the relationships between microphysical processes and hegemonic structural. They examine micropolitical forces, or what I have called the politics of death, and the way they were operationalized to crush grassroots resistance. The focus on the politics of death, allows for a capillary analysis of violence. It sheds light on micropolitical processes concerning urbicide in Aleppo. It examines the ways official nationalism was weaponized against Syrians, and their aspirations for democracy. The politics of life on the other hand, opens new vistas to comprehend Syrians' resistance against the devastating violence of the regime. A politics of life revolves around grassroots narratives and iterative experimentation.

This book is not a definitive assessment of revolutionary processes in Syria since 2011; it is rather an invitation to explore specific aspects of the revolt using unconventional theoretical tools and methodologies. It argues that the outcome of these unconventional tools is highly unpredictable. Unlike mainstream methodologies that constrain the purview of the research and predetermine the array of findings, some of the tools proposed here indicate new pathways. When geopolitics is the sole point of entry to examining the Syrian tragedy, then the starving bodies of Palestinians (and Syrians) living in besieged al-Yarmouk Camp are invisiblized; their agonizing voices are inaudible. International Relations, as a field of study, exists to maintain hegemonic relations of power. It is utilized to preserve the interests of the state and prevent non-state actors from disrupting the status quo. In the end, mainstream frameworks can become lethal in myriad ways, as the past eight years have amply shown us.

This book attempts to investigate the subterranean territories of the Syrian revolt. What one finds in these spaces are people struggling to create new institutions; negotiating ways to provide food for their communities; telling their stories despite monumental obstacles, at the same time they reinvent the meaning of the nation-state. They experiment with various tools to create livable spaces; and they do so without the guarantee of success. The iterative process is central in this context since revolutionaries are destroying the oppressive structures of the state and inventing new ways of being.

The Syrian revolution was developing strategies against authoritarian rule but equally as important, it was functioning in a world-system that speaks the language of the nation-state. The Syrian people were/are organizing against dictatorship but also struggling against an oppressive world order. For example, the United Nations refused to have any relationship with non-state actors and as a result refused to deliver food to the opposition controlled regions. The Syrian regime was tasked with distributing UN aid, while it was besieging dozens of regions and burning crops in insurgents areas. The UN provided medical equipment and medication to a state that had been targeting the medical infrastructure in a systematic way for years. The question then is how do people develop a critic of the state while, at the same time they recognize the real power of global institutions, laws, and economies in which their practices are rooted? Another challenge facing Syrians is the question of bottom-up deliberation and democracy. The technologies of violence and the politics of death deployed in Syria reached levels unmatched in recent history. The main perpetrator was obviously the Syrian regime but others had also their share. What kind of democratic praxes can be developed in a context that violence permeates? Finally, how do people develop grassroots strategies while at the same time operate within a world order that works against their aspirations?

Notes

INTRODUCTION

1. "Solidarity with the Syrian Revolution," *The Socialist Worker*, May 1, 2013. Accessed June 3, 2019. https://socialistworker.org/2013/05/01/solidarity-with-the-revolution.
2. Marcus Halaby, "Supporting Syria's Revolution at the Tunis World Social Forum," *Red Flag Online*, April 14, 2013. Accessed May 3, 2019. www.redflagonline.org/2013/04/supporting-syrias-revolution-at-the-tunis-world-social-forum/.
3. Chris York and Ewan Somerville, "Professor Piers Robinson Leaves Sheffield Uni Post After Accusations of Promoting Conspiracy Theories," *Huffington Post*, April 17, 2019. Accessed June 9, 2019. https://bit.ly/2EPoZSo.
4. For an extensive discussion about the impact of climate change on the Syrian revolt, see Jan Selbya, Omar S. Dahi, Christiane Fröhlich, and Mike Hulmee, "Climate change and the Syrian Civil War revisited," *Political Geography* 60 (September 2017): 232–244. https://doi.org/10.1016/j.polgeo.2017.05.007.
5. Francesca De Châtel, "The role of drought and climate change in the Syrian Uprising: untangling the triggers of the Revolution," *Middle Eastern Studies* 50 (2014): 521–535. DOI: 10.1080/00263206.2013.850076.
6. Annie Sparrow, "How a UN Health Agency became an apologist for Assad atrocities," *Middle East Eye*, January 16, 2017. Accessed June 19, 2019. www.middleeasteye.net/big-story/how-un-health-agency-became-apologist-assad-atrocities.
7. See Academic page of Theodore Postol at: https://sts-program.mit.edu/people/emeriti-faculty/theodore-postol/.
8. "MIT Professor accuses Bellingcat's Higgins of enabling war criminals to walk free in Syria (VIDEO)," *RT*, October 21, 2018. Accessed June 5, 2019. www.rt.com/news/441891-bellingcat-higgins-debate-syria/.
9. For an example of a rebuke to Theodore Postol, see Cheryl Rofer, "'A naive set of assumptions' – an expert's view on Ted Postol hexamine theories," *Bellingcat*, August 6, 2018. Accessed July 9, 2019. www.bellingcat.com/news/mena/2018/08/06/naive-set-assumptions-experts-view-ted-postol-hexamine-theories/; and Eliot Higgins, "Professor Theodore A. Postol of MIT vs. The Concept of Time," *Bellingcat*, April 28, 2017. Accessed July 9, 2019.www.bellingcat.com/resources/articles/2017/04/28/professor-theodore-postol-mit-vs-concept-time/.

10. For a discussion of the ways the Syrian regime instrumentalized Palmyra to polish its image, see Nour A. Munawar, "Reconstructing cultural heritage in conflict zones: should Palmyra be rebuilt?," EX NOVO *Journal of Archaeology* 2 (December 2017): 33–48.
11. Irina Bokova has a close relationship with the Russian autocrat and is more than willing to use UNESCO to enhance his image. See Sam Greenhill, "Putin Ally in UNESCO crony row is favourite to take top job at United Nations when Ban Ki-Moon steps down at the end of this year," *Daily Mail*, April 12, 2016. Accessed July 9, 2019. www.dailymail.co.uk/news/article-3536896/Putin-ally-UNESCO-crony-row-favourite-job-United-Nations-Ban-Ki-moon-steps-end-year.html.
12. "Telephone conversation with UNESCO Director-General Irina Bokova," *The Kremlin*, March 27, 2016. Accessed July 9, 2019. http://en.kremlin.ru/events/president/news/51574.
13. Ian Black, "'They are barbarians': meet the man defending Syria's heritage from Isis," *Guardian*, September 29, 2015. Accessed July 9, 2019. www.theguardian.com/world/2015/sep/29/they-are-barbarians-meet-the-man-maamoun-abdulkarim-defending-syrias-heritage-from-isis.
14. Ibid.
15. Ibid.
16. Agence France-Presse in Berlin, "Syrian troops looting ancient city Palmyra, says archaeologist," *Guardian*, June 1, 2016. Accessed June 2, 2019. www.theguardian.com/world/2016/jun/01/syrian-troops-looting-ancient-city-palmyra-says-archaeologist.
17. Steven Lee Myers and Nicholas Kulish, "'Broken system' allows ISIS to profit from looted antiquities," *New York Times*, January 9, 2016. Accessed June 1, 2019. www.nytimes.com/2016/01/10/world/europe/iraq-syria-antiquities-islamic-state.html; Benoit Faucon, Georgi Kantchev, and Alistair MacDonald, "The men who trade ISIS loot," *Wall Street Journal*, August 6, 2017. Accessed July 8, 2019. www.wsj.com/articles/the-men-who-trade-isis-loot-1502017200; and Simon Cox, "The men who smuggle the loot that funds IS," *BBC*, February 17, 2015. Accessed July 8, 2019. www.bbc.com/news/magazine-31485439.
18. A group of Syrian archeologists, with the help of Western colleagues, built a network to prevent the looting. They catalogue artifacts before they are destroyed or stolen, and preserve whatever they can. Since they document the looting of government loyalists, they are also their targets. For a discussion of their work, see Joe Parkinson, Ayla Albayrak, and Duncan Mavin, "Syrian 'Monuments Men' race to protect antiquities as looting bankrolls terror," *Wall Street Journal*, February 10, 2015. Accessed July 8, 2019. www.wsj.com/articles/syrian-monuments-men-race-to-protect-antiquities-as-looting-bankrolls-terror-1423615241.
19. Lorella Ventura, "The 'Arab Spring' and Orientalist stereotypes: the role of Orientalism in the narration of the revolts in the Arab world," *Interventions* 19, no. 2 (2016): 282–97. DOI: 10.1080/1369801X.2016.1231587.

20. Deborah P. Dixon, *Feminist Geopolitics: Material States* (London: Routledge, 2016).
21. Paul Routledge, "Anti-Geopolitics," in John Agnew, Katharyne Mitchell, and Gearóid Ó Tuathail (Eds.), *A Companion to Political Geography* (Oxford: Blackwell, 2003), 236–48.
22. In *Life As Politics*, Asef Bayat proposes a theoretical toolbox to analyze social phenomena outside the West. For example, his concepts of "social nonmovment" and "quiet encroachment" shed light on aspects of oppositional politics in the Arab world and Iran in ways that New Social Movement Theory does not; see Asef Bayat, *Life As Politics: How Ordinary People Change the Middle East* (Stanford, CA: Stanford University Press, 2009).
23. Friedrich Nietzsche, "On the uses and disadvantages of history for life," in Daniel Breazeale (Ed.), *Untimely Meditations* (Cambridge: Cambridge University Press, 1997).

CHAPTER 1

1. Kevin W. Martin, *Syria's Democratic Years: Citizens, Experts, and Media in the 1950s* (Bloomington, IN: Indiana University Press, 2015).
2. Raymond Hinnebusch, *Syria: Revolution from Above* (London: Routledge, 2001), 37–9.
3. Hinnebusch, *Syria*, 42.
4. Ibid.
5. Myriam Ababsa, "Fifty Years of state land distribution in the Syrian Jazira: Agrarian Reform, Agrarian Counter-Reform, and the Arab Belt Policy (1958–2008)," in Habib Ayeb Reem Saad (Ed.), *Agrarian Transformation in the Arab World Persistent and Emerging Challenges* (Cairo: American University in Cairo Press, 2013).
6. Hanna Batatu, *Syria's Peasantry, the Descendants of Its Lesser Rural Notables, and Their Politics* (Princeton, NJ: Princeton University Press, 1999), 198.
7. Ibid., 198–201.
8. Steven Heydemann, "Tracking the 'Arab Spring': Syria and the Future of Authoritarianism," *Journal of Democracy* 24, no. 4 (2013), 66.
9. Radwan Ziadeh, *The Years of Fear: The Forcibly Disappeared in Syria* (Washington, DC: Freedom House, 2014).
10. Adam Hanieh, *Lineages of Revolt: Issues of Contemporary Capitalism in the Middle East* (Chicago, IL: Haymarket Books, 2013); Omar S. Dahi and Yasser Munif, "Revolts in Syria: tracking the convergence between authoritarianism and neoliberalism," *Journal of Asian and African Studies* 47, no. 4 (2012): 323–32; Gilbert Achcar, *The People Want: A Radical Exploration of the Arab Uprising*, trans. G.M. Goshgarian (Berkeley, CA: University of California Press, 2013); Joseph Daher, *Syria after the Uprisings: The Political Economy of State Resilience*, forthcoming (London: Pluto Press, 2019); Bassam Haddad, *Business Networks in Syria: The*

Political Economy of Authoritarian Resilience Book (Stanford, CA: Stanford University Press, 2011); and Batatu, Syria's Peasantry (Princeton, NJ: Princeton University Press, 1999).
11. Leon T. Goldsmith, Cycle of Fear: Syria's Alawites in War and Peace (London: Hurst and Company, 2015).
12. Edward Ziter, Political Performance in Syria: From the Six-Day War to the Syrian Uprising (New York: Palgrave Macmillan, 2015).
13. Karim Atassi, The Strength of an Idea (Cambridge: Cambridge University Press, 2018).
14. Philip S. Khoury, Syria and the French Mandate: The Politics of Arab Nationalism, 1920–1945 (Princeton, NJ: Princeton University Press, 2014); 185.
15. Michael Provence, "French Mandate counterinsurgency and the repression of the Great Syrian Revolt," in Cyrus Schayegh and Andrew Arsan (Eds.), The Routledge Handbook of the History of the Middle East Mandates (Abingdon: Routledge, 2015), 136–51.
16. Atassi, The Strength of an Idea, 150.
17. Khaled Yacoub Oweis, "Assad ends state of emergency," Reuters. Last modified April 20, 2011. Accessed May 18, 2019. www.reuters.com/article/us-syria/syrias-assad-ends-state-of-emergency-idUSTRE72N2MC20110421.
18. TIMEP, "TIMEP Brief: Law No. 19 of 2012: Counter-Terrorism Law," The Tahrir Institute of Middle East Policy. Last modified January 9, 2019. Accessed May 18, 2019. https://timep.org/reports-briefings/timep-brief-law-no-19-of-2012-counter-terrorism-law/.
19. Ziter, Political Performance in Syria, 57.
20. "Syria: The permanent state of emergency—a breeding ground for torture report submitted to the Committee Against Torture in the context of the review of the Initial Periodic Report of the Syrian Arab Republic," Al-Karama. Last modified April 9, 2010. Accessed May 18, 2019. https://tbinternet.ohchr.org/Treaties/CAT/Shared%20Documents/SYR/INT_CAT_NGO_SYR_44_10098_E.pdf.
21. Neil Macfarquhar and Alan Cowell, "Syrians said to approve Charter as battles go on," New York Times. Last modified February 27, 2012. Accessed May 18, 2019. www.nytimes.com/2012/02/28/world/middleeast/syrian-violence-continues-as-west-dismisses-new-charter.html?.
22. "The Syrian Constitution—1973–2012," Carnegie Middle East Center. Last modified December 5, 2012. Accessed May 18, 2019. https://carnegie-mec.org/diwan/50255?lang=en.
23. Ibid.
24. Ibid.
25. Michael Macaulay, "Syria: the need to reform monitoring of state of emergency," 19–20. Thesis for International Human Rights Professor Esmeralda Thornhill. Lawyers Rights Watch Canada. Last modified December 2, 2005. Accessed May 18, 2019. www.lrwc.org/ws/wp-content/uploads/2012/03/Syria.StateofEmergency.Macaulay.Feb_..06.pdf.
26. "Syria: a brief review," Amnesty International, 1979.

27. Ziadeh, *The Years of Fear*, 23.
28. "Commission on Human Rights, Question of Human Rights and States of Emergency, CHR," December 1998/108, UN CHROR, 54th Session, UN Doc. E/CN.4/DEC/1998/108 (1998); Leandro Despouy, "The administration of justice and the human rights of detainees: question of human rights and states of emergency," UN ESC, 49th Session, UN Doc. E/CN.4/Sub.2/1997/19, (1997); Tom Hadden, "Human rights abuses and the protection of democracy during states of emergency," in Eugene Cotran and Adel Omar Sheri (Eds.), *Democracy, the Rule of Law, and Islam* (The Hague: Kluwer Law International, 1999), 111.
29. Macaulay, "Syria: the need to reform."
30. Amnesty International, "Syria: 41 years of the state of emergency—Amnesty International reiterates its concerns over a catalogue of human rights violations," March 2004. Accessed May 20, 2019. http://web.amnesty.org/library/index/ENGMDE240162004.
31. Macaulay, "Syria: the need to reform," 25–6.
32. Giorgio Agamben, *State of Exception* (Chicago, IL: University of Chicago Press, 2005), 2.
33. Carl Schmitt, *Political Theology* (Chicago, IL: University of Chicago Press, 2006).
34. Hans Kelsen and Carl Schmitt, *The Guardian of the Constitution: Hans Kelsen and Carl Schmitt on the Limits of Constitutional Law* (Cambridge: Cambridge University Press, 2015).
35. Yehouda Shenhav, "Imperialism, exceptionalism, and the contemporary world," in Marcelo Svirsky and Simone Bignall (Eds.), *Agamben and Colonialism* (Edinburgh: Edinburgh University Press, 2012), 19–20.
36. Schmitt, *Political Theology*.
37. Agamben, *State of Exception*, 86–7.
38. Sergei Prozorov, *Agamben and Politics: A Critical Introduction* (Edinburgh: Edinburgh University Press, 2014), 115.
39. Walter Benjamin, "Theses on the philosophy of history," trans. Harry Zohn. *Illuminations*. Hannah Arendt (Ed.) (New York: Schocken, 1968), 254.
40. Giorgio Agamben, *Homo Sacer: Sovereign Power and Bare Life* (Stanford, CA: Stanford University Press, 1998), 1.
41. Marcelo Svirsky and Simone Bignall (Eds.), *Agamben and Colonialism* (Edinburgh: Edinburgh University Press, 2012); and Achille Mbembe, "Necropolitics," *Public Culture* 15, no. 1 (2003): 11–40.
42. Jasbir K. Puar, *Terrorist Assemblages: Homonationalism in Queer Times* (Durham, NC: Duke University Press, 2007); Steven C. Caton and Bernardo Zacka, "Abu Ghraib, the security apparatus, and the performativity of power," *American Ethnologist* 37, no. 2 (2010): 203–11; and Derek Gregory, "Vanishing points: law, violence and exception in the global war prison," in Derek Gregory and Allan Pred (Eds.), *Violent Geographies: Fear, Terror and Political Violence* (New York: Routledge, 2006).

43. Nurhan Abujidi, "The Palestinian states of exception and Agamben," *Contemporary Arab Affairs* 2, no. 2 (2009): 272–91; Ronit Lentin, "Palestine/Israel and state criminality: exception, settler colonialism and racialization," *State Crime Journal* 5, no. 1 (2016): 32–50; Stephen Morton, "The Palestinian state of emergency and the art practice of Emily Jacir," in P. Lichtenfels and J. Rouse (Eds.), *Performance, Politics and Activism: Studies in International Performance* (London: Palgrave Macmillan, 2013); and Sari Hanafi and Taylor Long, "Governance, governmentalities, and the state of exception in the Palestinian refugee camps of Lebanon," *Journal of Refugee Studies* 23, no. 2 (2010): 134–59.
44. Lucia Ardovini and Simon Mabon, "Egypt's unbreakable curse: tracing the state of exception from Mubarak to Al Sisi," *Mediterranean Politics* (2019): 1–20, DOI: 10.1080/13629395.2019.1582170; Salwa Ismail, *The Rule of Violence: Subjectivity, Memory and Government in Syria* (Cambridge: Cambridge University Press, 2018); Jules Etjim, "Notes on Syria and the coming global thanatocracy," *Paths and Bridges*. Last modified July 11, 2018. Accessed May 19, 2019. https://pathsandbridges.wordpress.com/2018/07/11/notes-on-syria-and-the-coming-global-thanatocracy/amp/?__twitter_impression=true; George Abu Ahmad, "Order, freedom and chaos: sovereignties in Syria," *Middle East Policy* 2 (Summer 2013): 47–54; and Abdulhay Sayed, "In the Syrian prison: disconnected and desubjectified," *Global Dialogue* 4, no. 2 (2014). Accessed May 20, 2019.
45. Hinnebusch, *Syria*, 44–60.
46. See Nikolaos Van Dam, *The Struggle for Power in Syria: Politics and Society Under Asad and the Ba'th Party* (London: I.B. Tauris, 2011).
47. Carnegie Middle East Center, "Syrian Constitution 1973–2012."
48. Ziadeh, *The Years of Fear*, 12–14.
49. Raphaël Lefèvre, *Ashes of Hama: The Muslim Brotherhood in Syria* (Oxford: Oxford University Press, 2013), 76.
50. James A. Paul, *Syria Unmasked* (New York: Human Rights Watch. 1991).
51. "The Damascus Spring," *Carnegie Middle East Center*. Last Modified April 1, 2012. Accessed May 20, 2019. https://carnegie-mec.org/diwan/48516?lang=en.
52. Human Rights Watch, "A wasted decade: human rights in Syria during Bashar al-Asad's first ten years in power." Last Modified July 16, 2010. Accessed May 20, 2019. www.hrw.org/report/2010/07/16/wasted-decade/human-rights-syria-during-bashar-al-asads-first-ten-years-power.
53. Samer Abboud, *Syria*, 2nd edn (Cambridge: Polity, 2018), 56.
54. Atassi, *The Strength of an Idea*, 358–9.
55. Mukhabrat refers to intelligence branches or individuals. There are three security branches in Syria: General Intelligence, Military Intelligence, and Air Force Intelligence.
56. Lefèvre, *Ashes of Hama*, 158.
57. "Far from justice Syria's Supreme State Security Court," *Human Rights Watch*. Last modified February 24, 2009, Accessed May 19, 2019. www.

hrw.org/report/2009/02/24/far-justice/syrias-supreme-state-security-court.
58. Jane Mayer, "Outsourcing torture," *New Yorker*. Last modified February 6, 2005. Accessed May 19, 2019. www.newyorker.com/magazine/2005/02/14/outsourcing-torture.
59. "Syria: Counterterrorism Court used to stifle dissent," *Human Rights Watch*. Last modified June 25, 2013. Accessed May 19, 2019. www.hrw.org/news/2013/06/25/syria-counterterrorism-court-used-stifle-dissent.
60. In *Life as Politics: How Ordinary People Change the Middle East* (Stanford, CA: Stanford University Press, 2009), Asef Bayat proposes a complex framework to examine politics in the Middle East without reproducing Orientalist tropes. He examines the various ways people organize and resist without necessarily using the language or the tools of political parties or social movements in Western countries.
61. Mbembe, "Necropolitics," 18.
62. Ibid., 12.
63. Ibid., 21.
64. Ibid., 40.
65. Ibid., 23.
66. Aimé Césaire, *Discourse on Colonialism* (New York: Monthly Review Press, 1972), 35–6.
67. Mbembe, "Necropolitics," 21.
68. The Syrian regime caused most of the destruction and killing, but other actors were involved as well. They include Russia, Iran, ISIS, the United States, and rebel groups. See this Human Rights Watch report: "Syria: events of 2017," *Human Rights Watch*. Accessed May 19, 2019. www.hrw.org/world-report/2018/country-chapters/syria.
69. Amal Alachkar, "A Syrian scholar in exile," *News Deeply*. Last modified October 8, 2013. Accessed May 19, 2019. www.newsdeeply.com/syria/community/2013/10/08/a-syrian-scholar-in-exile.
70. "Human rights situations that require the Council's attention," Human Rights Council Thirty-Seventh Session 26 February–23 March 2018, Agenda item 4. Conference room paper of the Independent International Commission of Inquiry on the Syrian Arab Republic. Accessed May 20, 2019. www.ohchr.org/Documents/HRBodies/HRCouncil/CoISyria/A-HRC-37-CRP-3.pdf.
71. Raluca Albu and Raad Rahman, "Leaving Aleppo: crossing Syria's most dangerous checkpoints," *The Rumpus*. Last modified September 19, 2016. Accessed May 20, 2019. https://therumpus.net/2016/09/leaving-aleppo-crossing-syrias-most-dangerous-checkpoints/.
72. David Kilcullen, Nate Rosenblatt, and Jwanah Qudsi, "Mapping the conflict in Syria: Aleppo, Syria," *Caerus*. Last modified February 2014. Accessed May 20, 1019. http://caerusassociates.com/wp-content/uploads/2014/02/Caerus_AleppoMappingProject_FinalReport_02-18-14.pdf.
73. Lava Selo, "The deadly checkpoint that divides Syria's biggest city," *NPR*. September 6, 2013. Accessed May 20, 1019. www.npr.org/sections/

parallels/2013/09/05/219258334/the-deadly-checkpoint-that-divides-syrias-biggest-city.
74. Salim Salameh, "Starving the Palestinian Yarmouk Camp," *Carnegie Middle East Center*. Last modified April 28, 2014. Accessed May 20, 1019. https://carnegie-mec.org/diwan/55450.
75. Zabadani, which is located in the west near the Lebanese border, was the first to organize a democratic election and elect a revolutionary council. The city was besieged and starved until it surrendered, and its inhabitants were forcefully displaced. With a few exceptions, the regime prevented food and medication from entering the city. See Anne Barnard and Hwaida Saas, "In Syrian town cut off from the world, glimpses of deprivation," *New York Times*. Last modified January 14, 2016. Accessed May 20, 2019. www.nytimes.com/2016/01/15/world/middleeast/madaya-syria.html.
76. Sadik Abdul Rahman, "Maarrat al-Nu'man: a hundred days of confrontation with al-Nusra Front," *Al-Jumhuriya*. Last modified June 21, 2016. Accessed May 20, 2019. www.aljumhuriya.net/en/content/maarrat-al-nu'man-hundred-days-confrontation-al-nusra-front.
77. Barbara Walters, "Transcript: ABC's Barbara Walters' interview with Syrian President Bashar al-Assad," *ABC News*. Last modified December 7, 2011. Accessed May 20, 2019. https://abcnews.go.com/International/transcript-abcs-barbara-walters-interview-syrian-president-bashar/story?id=15099152.
78. "Syria conflict: air strikes on Idlib markets 'kill dozens,'" *BBC World News*. Last modified April 19, 2016. Accessed May 20, 2019. www.bbc.com/news/world-middle-east-36084848.
79. Ellen Francis, "The war on Syria's doctors," *Foreign Policy*. Last modified August 11, 2016. Accessed May 20, 2019. https://foreignpolicy.com/2016/08/11/the-war-on-syrias-doctors-assad-medicine-underground/; and Peter Beaumont, "Shoot the journalists: Syria's lesson from the Arab spring," *The Guardian*. Last modified February 25, 2012. Accessed May 20, 2019. www.theguardian.com/world/2012/feb/26/syria-targets-journalists.
80. David Butter, "Syria's economy picking up the pieces," *Chatham House*. Last modified June 2015. Accessed May 20, 2019. www.chathamhouse.org/sites/default/files/field/field_document/20150623SyriaEconomyButter.pdf.
81. Raymond A. Hinnebusch, *Authoritarian Power and State Formation in Ba'thist Syria* (Boulder, CO: Westview Press, 1990); Hanna Batatu, "Syria's Muslim Brethren," in Fred Halliday and Hamza Alavi (Eds.), *State and Ideology in the Middle East and Pakistan* (London: Palgrave, 1988), 112–32; Batatu, *Syria's Peasantry*; and Fred H. Lawson, "Social bases for the Hamah Revolt," *MERIP*. Last modified November/December 1982. Accessed May 20, 2019. https://merip.org/1982/11/social-bases-for-the-hama-revolt/.
82. "The Syrian uprising: the balance of power is shifting," *Economist*. Last modified June 9, 2011. Accessed May 20, 2019. www.economist.com/middle-east-and-africa/2011/06/09/the-balance-of-power-is-shifting.

83. Yassin al-Hajj Saleh, "The Syrian Shabiha and their state," *Heinrich Böll Stiftung*. Last modified April 16, 2012. Accessed May 20, 2019. https://lb.boell.org/en/2012/04/16/syrian-shabiha-and-their-state.
84. "US will not intervene in Syria as it has in Libya, says Hillary Clinton," *Guardian*. Last modified March 27, 2011. Accessed May 20, 2019. www.theguardian.com/world/2011/mar/27/report-12-killed-syrian-port-city.
85. "Death of Druze leader reported in Syria blast," *Al-Jazeera*. Last modified September 5, 2015. Accessed May 20, 2019. www.aljazeera.com/news/2015/09/death-druze-leader-reported-syria-blast-150905022304903.html.
86. "Syria's Druze minority: walking a war-time tightrope," *France 24*. Last modified July 30, 2018. Accessed May 20, 2019. www.france24.com/en/20180730-syrias-druze-minority-walking-war-time-tightrope.
87. Anne Speckhard and Ardian Shajkovci, "After a new massacre, charges that ISIS is operating with Assad and the Russians," *Daily Beast*. Last modified August 9, 2018. Accessed May 20, 2019. www.thedailybeast.com/how-assad-isis-and-the-russians-cooperated-to-carry-out-a-massacre.
88. Anthony Shadid, "Killing of opposition leader in Syria provokes Kurds," *New York Times*. Last modified October 8, 2011. Accessed May 20, 2019. www.nytimes.com/2011/10/09/world/middleeast/killing-of-opposition-leader-in-syria-provokes-kurds.html.
89. "Statement by the Kurdish Youth Movement (TCK) about the latest events in the city of Amouda, and videos and pictures from the protests and sit ins," *Syria Freedom Forever*. Last modified June 23, 2013. Accessed May 20, 2019. https://syriafreedomforever.wordpress.com/2013/06/23/statement-by-the-kurdish-youth-movement-tck-about-the-latest-events-in-the-city-of-amouda-and-videos-and-pictures-from-the-protests-and-sit-ins/.
90. Paulo Gabriel Hilu Pinto, "The shattered nation: the sectarianization of the Syrian Conflict," in Nader Hashemi and Danny Postel (Eds.), *Sectarianization: Mapping the New Politics of the Middle East* (Oxford: Oxford University Press, 2017).
91. Matthias Sulz, "Loyalty over geography: re-interpreting the notion of 'Useful Syria,'" *Syria Comment*. Last modified September 6, 2018. Accessed May 20, 2019. www.joshualandis.com/blog/loyalty-over-geography-re-interpreting-the-notion-of-useful-syria-by-matthias-sulz/.
92. "Table of besieged communities in Syria from upcoming Siege Watch Report," *Siege Watch*. Last modified October 2015. Accessed May 20, 2019. https://siegewatch.org/wp-content/uploads/2015/10/SiegeWatchTable-Final-for-release.pdf.
93. "Under siege: the plight of East Ghouta," *Syrian American Medical Society*. Last modified September 2017. Accessed May 20, 2019. www.sams-usa.net/wp-content/uploads/2017/06/east-ghouta-report-06.pdf.
94. "Slow death: life and death in Syrian communities under siege," *Syrian American Medical Society*. Last modified March 2015. Accessed May 20,

2019. www.sams-usa.net/wp-content/uploads/2016/09/Slow-Death_Syria-Under-Siege.pdf.
95. Annie Sparrow, "How a UN health agency became an apologist for Assad atrocities," *Middle East Eyes*. Last modified January 16, 2017. Accessed May 20, 2019. www.middleeasteye.net/big-story/how-un-health-agency-became-apologist-assad-atrocities.
96. Annie Sparrow, "How UN Humanitarian Aid has propped up Assad: Syria shows the need for reform," *Foreign Affairs*. Last modified September 20, 2018. Accessed May 20, 2019. www.foreignaffairs.com/articles/syria/2018-09-20/how-un-humanitarian-aid-has-propped-assad.
97. Bethany Allen-Ebrahimian, "Hospitals become the front line in the Syrian Civil War," *Foreign Policy*. Last modified May 31, 2017. Accessed May 20, 2019.https://foreignpolicy.com/2017/05/31/syria-hospitals-assad-civil-war-russia-usaid/.
98. Syria TV, "The necks of the neck and the neck ..." Jalal Mando, witness to the murders ... O Freedom (English Subtitles). *YouTube*. Video File. August 2, 2018. Accessed May 20, 2019. www.youtube.com/watch?v=E9_ccYOWqlo&frags=pl%2Cwn.
99. For an in-depth discussion, see Abdulhay Sayed, "In the Syrian prison: disconnected and desubjectified," *Global Dialogue* 4, no. 2 (2014). Accessed May20,2019.http://globaldialogue.isa-sociology.org/in-the-syrian-prison-disconnected-and-desubjectified/.
100. Syria TV, "The necks of the neck and the neck ...,".
101. Lizzie Porter, "Former detainees recount torture, organ harvesting in Syria's prisons," *Middle East Eyes*. Last modified 13 August 2016. Accessed May 20, 2019. www.middleeasteye.net/news/former-detainees-recount-torture-organ-harvesting-syrias-prisons.
102. Tohama Marouf, "Tihameh known: credit for the children of Daraa out of jail in 2011 YAHRIYA," (Arabic). *YouTube*. Video File. *Syria TV*. May 17, 2018. Accessed May 20, 2019. www.youtube.com/watch?v=aLCzolIUfo4&list=PLeMwite1QcQ3JIAdAEsJ_8ySjbjfoweai&index=34&frags=pl%2Cwn.
103. Louisa Loveluck and Zakaria Zakaria, "'The hospitals were slaughterhouses': a journey into Syria's secret torture wards," *Washington Post*. April 2, 2017. Accessed May 20, 2019. www.washingtonpost.com/world/middle_east/the-hospitals-were-slaughterhouses-a-journey-intosyrias-secret-torture-wards/2017/04/02/90ccaa6e-0d61-11e7-b2bb-417e331877d9_story.html?utm_term=.bc0afodaa5b4.
104. Christoph Koettl, "UN reveals further evidence of atrocities in Syria," *Amnesty International USA*. Last modified August 15, 2012. Accessed May 20,2019.https://blog.amnestyusa.org/middle-east/un-reveals-further-evidence-of-atrocities-in-syria/.
105. Mbembe, "Necropolitics," 28.
106. Maksymilian Czuperski, Faysal Itani, Ben Nimmo, Eliot Higgins, and Emma Beals, "Breaking Aleppo: Hospital Attacks," *Atlantic Council*. Last

modified February 2017. Accessed May 20, 2019. www.publications.atlanticcouncil.org/breakingaleppo/hospital-attacks/.
107. Eliot Higgins, Emma Beals, Ben Nimmo, Faysal Itani, and Maks Czuperski, "Breaking Aleppo," *Atlantic Council*. Last modified February 2017. Accessed May 20, 2019. www.publications.atlanticcouncil.org/breakingaleppo/.
108. Martin Chulov and Mona Mahmood, 'Syrian rebels recover scores of bodies from Aleppo River as floodwaters recede," *Guardian*. Last modified January 29, 2013. www.theguardian.com/world/2013/jan/29/syrian-rebels-bodies-aleppo-canal.
109. Agamben, *State of Exception*, 23.
110. 0areejo, "Scandal of Al Dunya TV anchor in Daraya 26/8/2012," (Arabic). *YouTube*. Video File. August 26, 2012. Accessed May 20, 2019. www.youtube.com/watch?v=3gCeE5e2k5g.
111. Janine di Giovanni, "Syria crisis: Daraya massacre leaves a ghost town still counting its dead," *Guardian*. Last modified September 7, 2012. Accessed May 20, 2019. www.theguardian.com/world/2012/sep/07/syria-daraya-massacre-ghost-town.
112. Chip Carey, "Syria's Civil War has become a genocide September," *World Policy* 16 (2013). https://worldpolicy.org/2013/09/16/syrias-civil-war-has-become-a-genocide/.
113. A *takfiri* is a Muslim who accuses another Muslim of apostasy. While al-Qaeda and ISIS are takfiri groups, the large majority of the Islamist opposition rejected takfirism. The regime branded the entire opposition (Islamist and secular) of takfirism to legitimize its war against Sunnis who opposed Assad.
114. "Record of arbitrary arrests," *Syrian Network for Human Rights*. September 24, 2018. Accessed May 20, 2019. http://sn4hr.org/blog/2018/09/24/record-of-arbitrary-arrests1/.
115. Mustafa Khalifeh, Interview with the author on Syria TV.
116. Joyce Laverty Miller, "The Syrian Revolt of 1925," *International Journal of Middle East Studies* (1977): 550–5.
117. Cited in Miriam Cooke, "The cell story: Syrian prison stories after Hafiz Asad," *Middle East Critique* 20, no. 2 (2011): 169–84.
118. Abdulhay Sayed, "In the Syrian prison: disconnected and desubjectified," *Global Dialogue* 4, no. 2 (2014). Accessed May 20, 2019.
119. Cooke, "The cell story," 184.
120. Laleh Khalili and Jillian Schwedler, "Introduction," in *Policing and Prisons in the Middle East—Formations of Coercion* (Oxford: Oxford University Press, 2010), 22–3.
121. Sune Haugbolle, "The victim's tale in Syria: imprisonment, individualism, and liberalism," in Laleh Khalili and Jillian Schwedler (Eds.), *Policing and Prisons in the Middle East—Formations of Coercion* (Oxford: Oxford University Press, 2010).

122. Syria TV, "O Freedom Mustafa Khalifa: a witness on the Syrian shell that begins with cells," *YouTube*. Video File. April 12, 2018. Accessed May 20, 2019. www.youtube.com/watch?v=weeIqvOS-xE.
123. Syria TV, "O Freedom Mustafa Khalifa."
124. Yassin Haj Saleh, *Bil Ikhlas ya Shabab* (London: Dar al-Saqi, 2017), 110–12.
125. Most scholars who work on the prison system in the United States highlight primarily its economic dimension. Some academics have pushed back against the economistic framework. The work of Laura Whitehorn shows the importance of the non-economic dimension in the prison–industrial complex (PIC). She argues that the PIC is a form of genocide of black people, not simply a source of revenue for big corporations. See Laura Whitehorn, "Black Power incarcerated: political prisoners, genocide, and the state," *Socialism and Democracy* 28, no. 3 (2014): 101–17, DOI: 10.1080/08854300.2014.954928.
126. Ruth Wilson Gilmore, *Golden Gulag: Prisons, Surplus, Crisis, and Opposition in Globalizing California* (Berkeley, CA: University of California Press, 2007); and Elizabeth Hinton, *From the War on Poverty to the War on Crime: The Making of Mass Incarceration in America* (Cambridge, MA: Harvard University Press, 2017).
127. Yassin Haj Saleh, "*Alsunh altdmuryh: sydnaya, althwl al'ensry, alebadh*," March 24, 2017. Accessed May 20, 2019. *Al-Jumhuriya*. www.aljumhuriya.net/ar/37459.
128. Prozorov, *Agamben and Politics*, 102.

CHAPTER 2

1. Keith David Watenpaugh, *Being Modern in the Middle East: Revolution, Nationalism, Colonialism in the Arab Middle East* (Princeton, NJ: Princeton University Press, 2006), 32.
2. Bruce Masters, "The political economy of Aleppo in an age of Ottoman reform," *Journal of the Economic and Social History of the Orient* 53, no. 1/2 (2010): 290–316.
3. Ross Burns, *Aleppo: A History* (New York: Routledge, 2017), 256.
4. Masters, "The political economy of Aleppo," 293.
5. Burns, *Aleppo*, 257.
6. Bruce Masters, "The 1850 Events in Aleppo: an aftershock of Syria's incorporation into the capitalist world system," *International Journal of Middle East Studies* 22 (1990), 8.
7. Burns, *Aleppo*, 257.
8. Masters, "The political economy of Aleppo," 300.
9. Nora Lafi, "Building and destroying authenticity in Aleppo: heritage between conservation, transformation, destruction, and re-invention," in Christoph Bernhardt, Martin Sabrow, and Achim Saupe (Eds.), *Gebaute Geschichte: Historische Authentizität im Stadtraum* (Göttingen: Wallstein, 2017), 206–28, www.wallstein-verlag.de/9783835330139-gebaute-geschichte.html.

10. Watenpaugh, *Being Modern in the Middle East*, 48.
11. Lafi, "Building and destroying authenticity in Aleppo," 206–28.
12. Patrick Seale, *Asad: The Struggle for the Middle East* (Oakland, CA: University of California Press, 1989), 450.
13. Watenpaugh, *Being Modern in the Middle East*, 48.
14. As cited in Daniel Neep, *Occupying Syria under the French Mandate* (Cambridge: Cambridge University Press, 2012), 150–1.
15. Lafi, "Building and destroying authenticity in Aleppo," 206–28.
16. André Raymond, "Islamic city, Arab city: Orientalist myths and recent views," *British Journal of Middle Eastern Studies* 21, no. 1 (1994), 5.
17. Neep, *Occupying Syria*, 102.
18. Raymond, "Islamic city, Arab city," 5.
19. Lafi, "Building and destroying authenticity in Aleppo," 206–28.
20. Kosuke Matsubara, "Gyoji Banshoya (1930–1998): a Japanese planner devoted to historic cities in the Middle East and North Africa," *Planning Perspectives* 31, no. 3 (2016): 391–423.
21. Ibid., 411.
22. Ibid., 212–14.
23. Hanna Batatu, *Syria's Peasantry, the Descendants of Its Lesser Rural Notables, and Their Politics* (Princeton, NJ: Princeton University Press, 1999).
24. Hanna Batatu, "Syria's Muslim Brethren," in Fred Halliday and Hamza Alavi (Eds.), *State and Ideology in the Middle East and Pakistan* (London: Palgrave, 1988).
25. Tine Gade, "Together all the way? abeyance and co-optation of Sunni networks in Lebanon," *Social Movement Studies* 18, no. 1 (2019): 56–77.
26. James A. Paul, *Human Rights in Syria* (New York: Middle East Watch, 1990), 18.
27. Jwanah Qudsi, "Rebuilding Old Aleppo: postwar sustainable recovery and urban refugee resettlement". Master of Urban Planning Thesis, New York University, 2017. www.academia.edu/26963608/Rebuilding_Old_Aleppo_Postwar_Sustainable_Recovery_and_Urban_Refugee_Resettlement.
28. Wajiha Mouhana, "The regime and revolution, dialectics of city and countryside: Aleppo as a case study," *As-safir*. Beirut. February 20, 2013. http://assafirarabi.com/ar/3326/2013/02/20/جدلية-الاريف-والمدينة-الاثورة-والسلطة/.
29. Ibid.
30. Bara Halabi, "Dignity Strike: the old market—Skayta, 2-6-2012," *YouTube*. Video file. Posted June 3, 2012. Accessed June 2, 2019. www.youtube.com/watch?v=voF4fBY2tjQ; and Syrian Revolution Documentation Center, "Aleppo, Old Manshiyya—General Strike 16-6-2012 (Arabic)," *YouTube*. Video file. Posted June 16, 2012. Accessed June 2, 2019. www.youtube.com/watch?v=ZBFd9FV4qdg.

31. Saber Darwīsh and Mohammad Abī Samrā, *ma'si halb: al-thourah al-maghdourah wa rasaa'l al-muhasarin* [*Tragedies of Aleppo: The Betrayed Revolution and Messages of the Besieged*] (Milan: al-Mutawassit, 2016), 42.
32. Rami Makhlouf is the cousin of Bashar al-Assad, and the wealthiest person in Syria. He was part of Assad's inner circle, and played a central role repressing the revolt by funding militias. Protesters throughout Syria created songs and slogans to denounce his corruption and repressive role in the revolt.
33. Mouhana, "The Regime and Revolution."
34. FSA1operations, "Aleppo—al-Sukkari District. The Free Army facing Shabih (Arabic)." *YouTube*. Video file. Last modified March 23, 2012. Accessed June 2, 2019. www.youtube.com/watch?v=iVT8BckCdZA.
35. Nurhan Abujidi, *Urbicide in Palestine: Spaces of Oppression and Resilience* (London: Routledge, 2014), 29.
36. Ibid., 30–2.
37. Ibid., 32.
38. Ibid., 33.
39. Ibid., 40–2.
40. A Takfiri is a Muslim who accuses another Muslim of apostasy.
41. The regime claimed that Syrian Islamists were paid by the Wahhabist Saudi family to oppose Assad. The Syrian government argued that protesters were manipulated by Saudi security to destabilize the Syrian state, and as such undermine the axis of resistance, which includes Syria, Iran, Palestinian groups, and Hezbollah.
42. Phil Sands, Justin Vela, and Suha Maayeh, "Assad regime abetted extremists to subvert peaceful uprising, says former intelligence official," *The National*. Last modified January 14, 2014. Accessed June 2, 2019. www.thenational.ae/world/assad-regime-abetted-extremists-to-subvert-peaceful-uprising-says-former-intelligence-official-1.319620.
43. Roy Gutman, "Assad henchman: here's how we built ISIS," *The Daily Beast*. Last modified January 12, 2016. Accessed June 2, 2019. www.thedailybeast.com/assad-henchman-heres-how-we-built-isis.
44. Richard Spencer, "Four jihadists, one prison: all released by Assad and all now dead," *The Telegraph*. Posted May 11, 2016. Accessed June 2, 2019. http://s.telegraph.co.uk/graphics/projects/isis-jihad-syria-assad-islamic/index.html.
45. "Violations against media activists," *Syrian Network for Human Rights*. Accessed October 5, 2019. http://sn4hr.org/public_html/wp-content/pdf/english/Violations%20against%20media%20activists.pdf. Sam Daher, "A cruel epilogue to the Syrian Civil War," *The Atlantic*. August 15, 2018. Accessed October 5, 2019. www.theatlantic.com/international/archive/2018/08/lists-of-dead-in-syria-assad/567559/.
46. Darwīsh and Abī Samrā, *ma'si halb* [*Tragedies of Aleppo*], 50.
47. "Has the regime begun to change the demographic composition of the eastern districts of Aleppo?" *Zaman alwasl*. Last modified November 11, 2018. Accessed June 2, 2019. www.zamanalwsl.net/news/article/96561.

48. The Crisis Group, *Syria's Metastasizing Conflicts*, 9. Last modified June 7, 2013. Accessed June 2, 2019. www.crisisgroup.org/middle-east-north-africa/eastern-mediterranean/syria/196-lessons-syrian-states-return-south.
49. Hadeel Al Shalchi and Erika Solomon, "Syrian soldier executed after graveside 'trial,'" *Reuters*. Last Modified August 1, 2012. Accessed June 2, 2019.www.reuters.com/article/us-syria-crisis-justice/syrian-soldier-executed-after-graveside-trial-idUSBRE8700KT20120801.
50. Regine Schwab, "Insurgent courts in civil wars: the three pathways of (trans)formation in today's Syria (2012–2017)," *Small Wars & Insurgencies* 29, no. 4 (2018): 801–26. DOI: 10.1080/09592318.2018.1497290.
51. Miriyam Aouragh, "Online and offline maneuverings in Syria's counter-revolution," *Jadaliyya*. Late modified September 5, 2014. Accessed June 2, 2019.www.jadaliyya.com/Details/30870/Online-and-Offline-Maneuverings-in-Syria%60s-Counter-Revolution.
52. Ausama Monajed, "The opposition position," *Foreign Policy*. Last modified October 3, 2011. Accessed June 2, 2019. https://foreignpolicy.com/2011/10/03/the-opposition-position/.
53. Jamal Shahid, "Rosa Yassin Hassan and the Masons," *Alaraby*. Last modified November 23, 2016. Accessed June 2, 2019. www.alaraby.co.uk/diffah/books/2016/11/23/روزا-ياسين-حسن-والماسون.
54. Clare M. Gillis, "Syria's Christians wary of both rebels and Assad regime," *Daily Beast*. Last modified October 17, 2012. Accessed June 2, 2019. www.thedailybeast.com/syrias-christians-wary-of-both-rebels-and-assad-regime?ref=scroll.
55. Helena Smith, "Syrian bishops kidnapped in Aleppo still missing one month on," *Guardian*. Last modified May 21, 2013. Accessed June 2, 2019. www.theguardian.com/world/2013/may/21/syrian-bishops-kidnapped-yazigi-ibrahim.
56. "Plea to free Archbishop Mar Gregorios Yohanna Ibrahim and Archbishop Boulos Yazigi who were kidnapped one year ago today," *Huffington Post*. Last Modified June 21, 2014. Accessed June 2, 2019. www.huffpost.com/entry/archbishop-mar-gregorios-yohanna-ibrahim-and-archbishop-boulos-yazigi_b_5186945.
57. Andrew Weaver, "Syria crisis: Aleppo bishops kidnapped—Tuesday April 23," *Guardian*. Last modified April 23, 2013. Accessed June 2, 2019. www.theguardian.com/world/middle-east-live/2013/apr/23/syria-bishops-kidnapped-rebels-live.
58. George K. Mayala, Mohammed Hammam, and Ziada L. Salen, "Aleppo: people, place, and war: Armenians in the Syrian crisis," *Suwar Magazine*. Accessed June 2, 2019. www.suwar-magazine.org/details/الناس-والحرب:-المكان-:حلب/293/category_health.html.
59. Ara Sanjian, "Armenians in the midst of civil wars: Lebanon and Syria compared," 2015, 1–25. Accessed June 2, 2019. www.academia.edu/17939611/Armenians_in_the_Midst_of_Civil_Wars_Lebanon_and_Syria_Compared_in_English_

60. Ibid., 21.
61. Arsen Hakobyan, "From Aleppo to Yerevan: the war and migration from the window to the bus," in Marcello Mollica (Ed.), *Fundamentalism: Ethnographies on Minorities, Discrimination and Transnationalism* (Berlin: LIT Verlag, 2016), 22.
62. Mikayel Zolyan, "Refugees or repatriates? Syrian Armenians return to Armenia," *Open Democracy*. Last modified October 15, 2015. Accessed June 2, 2019. www.opendemocracy.net/en/odr/refugees-or-repatriates-syrian-armenians-return-to-armenia/.
63. Stanimir Dobrev, "The upward mobility of Aleppo's Palestinians," *International Review*. Last modified January 28, 2018. Accessed June 2, 2019. https://international-review.org/upward-mobility-aleppos-palestinians/.
64. Ruth Sherlock, "Syria: Christians take up arms for the first time," *The Telegraph*. Last modified September 12, 2012. Accessed June 2, 2019. www.telegraph.co.uk/news/worldnews/middleeast/syria/9539244/Syria-Christians-take-up-arms-for-first-time.html.
65. Leith Abu Feidel, "Where the defense is kept by the Armenian militia, the IS fighters cannot advance an inch," *Rusarminfo*. Last modified September 19, 2015. Accessed June 2, 2019. https://rusarminfo.ru/2015/09/19/leith-abu-fadel-where-the-defense-is-kept-by-the-armenian-militia-the-is-fighters-cannot-advance-an-inch/.
66. Aleppo Project 2012.
67. The much feared Syrian security forces.
68. Alaraby TV, "I was there: Aleppo University bombing," *YouTube*. March 8, 2018. Accessed October 5, 2019. www.youtube.com/watch?v=K8FGNcdhPWg.
69. Mayala, Hammam, and Salen, "Aleppo: people, place, and war."
70. "University protest in Saria 'turns deadly,'" *Al Jazeera*. Last modified May 3, 2012. Accessed June 2, 2019. www.aljazeera.com/news/middleeast/2012/05/201253841824918o.html.
71. "Syria: 'Heroes of Aleppo University' protests-Friday May 18," *Guardian*. Last modified May 18, 2012. Accessed June 2, 2019. www.theguardian.com/world/middle-east-live/2012/may/18/syria-aleppo-university-protests-live.
72. Joseph Daher, "The Student Movement in Syria," *Syria Frontline* blog (Arabic). Last modified January 24, 2013. Accessed June 2, 2019. http://syria.frontline.left.over-blog.com/article-114701217.html.
73. Malak Chabkoun, "Pro-regime militias in Syria: SAA Unit or ad-hoc apparatus?" Al Jazeera. Last modified August 4, 2014. Accessed June 2, 2019. http://studies.aljazeera.net/en/reports/2014/07/201472494759578879.html.
74. Haian Dukhan, "Tribes and tribalism in the Syrian Uprising," Syria Studies 6, no. 2 (2014): 4–5.
75. Ibid., 5.
76. Brigit Schaebler, "Constructing an identity between Arabism and Islam: The Druzes in Syria," *The Muslim World* 103, no. 1 (2013): 76–7.

77. Wladimir van Wilgenburg, "Syrian Kurds appoint Arab governor in Hasakah, bid for international support," *Middle East Eye*. Last modified July 31, 2014. Accessed June 2, 2019. www.middleeasteye.net/news/syrian-kurds-appoint-arab-governor-hasakah-bid-international-support.
78. BSYRIA, "Will Aleppo rise up?" *Foreign Policy*. Last modified February 21, 2012. Accessed June 2, 2019. https://foreignpolicy.com/2012/02/21/will-aleppo-rise-up/.
79. "Revolution and battles in Aleppo," *Syria Freedom Forever*. Last modified August 17, 2014. Accessed June 2, 2019. https://syriafreedomforever.wordpress.com/2014/08/17/الثورة-والمعارك-في-حلب/.
80. "'Berri' family, history of shame in Aleppo," *Zaman Alwasl*. Last modified July 25, 2012. Accessed June 2, 2019. https://en.zamanalwsl.net/news/article/854/
81. "Syria: the silenced Kurds," *Human Rights Watch* 8, no. 4E (1996). www.hrw.org/reports/1996/Syria.htm#P229_37096.
82. David McDowall, *A Modern History of the Kurds*, 3rd edn (London: I.B. Tauris, 2007), 476.
83. Angus McDowall, "Aleppo district shows Assad's delicate dance with Kurds," *Reuters*. Last modified July 27, 2017. Accessed June 2, 2019. www.reuters.com/article/us-mideast-crisis-aleppo-kurds/aleppo-district-shows-assads-delicate-dance-with-kurds-idUSKBN1AC1SY.
84. John Caves, "Syrian Kurds and the Democratic Union Party (PYD)," *Understanding War*. Last modified December 6, 2012. Accessed June 2, 2019. www.understandingwar.org/sites/default/files/Backgrounder_Syrian Kurds.pdf.
85. Sebastian Gonano, "The role of Sheikh Maqsoud during the battle for Aleppo," *International Review*. Last modified November 16, 2017. Accessed June 2, 2019. https://international-review.org/the-role-of-sheikh-maqsoud-during-the-battle-for-aleppo/.
86. United Nations, A/HRC/34/64 General Assembly Human Rights Council Thirty-fourth session 27 February–24 March 2017. "Report of the Independent International Commission of Inquiry on the Syrian Arab Republic."
87. "Syria: armed opposition groups committing war crimes in Aleppo city," *Amnesty International*. Last modified May 13, 2016. Accessed June 2, 2019. www.amnesty.org/en/latest/news/2016/05/syria-armed-opposition-groups-committing-war-crimes-in-aleppo-city/.
88. Arwa Ibrahim, "Analysis: the Kurdish 'frenemies' aiding Assad in Aleppo," *Middle East Eyes*. Last modified November 30, 2016. Accessed June 2, 2019.www.middleeasteye.net/news/analysis-kurdish-frenemies-aiding-assad-aleppo.
89. "Group denial: repression of Kurdish political and cultural rights in Syria," *Human Rights Watch*. Last modified November 26, 2009. Accessed June 2, 2019. www.hrw.org/report/2009/11/26/group-denial/repression-kurdish-political-and-cultural-rights-syria.

90. Robert Lowe, *The Syrian Kurds: A People Discovered* (London: Chatham House, 2006), www.chathamhouse.org/sites/default/files/public/Research/Middle%20East/bpsyriankurds.pdf.
91. Alex de Jong, "The Rojava Project," *Jacobin*. Last modified November 30, 2016. Accessed June 2, 2019. https://jacobinmag.com/2016/11/rojava-syria-kurds-ypg-pkk-ocalan-turkey.
92. "Moukhayam al Nayrab (Nayrab Camp)," *Wafa*. Accessed June 2, 2019. www.wafainfo.ps/ar_page.aspx?id=8853.
93. "Aleppo's War of the Peripheries," *Alaalam*. Last modified October 25, 2016. Accessed June 2, 2019. http://alaalam.org/ar/society-and-culture-ar/item/419-598251016.
94. Michael Karadjis, "Syria and the Palestinians: no other Arab state has as much Palestinian blood on its hands," *Syrian Revolution Commentary and Analysis* blog. Last modified September 24, 2013. Accessed June 2, 2019. https://mkaradjis.wordpress.com/2013/09/24/syria-and-the-palestinians-almost-no-other-arab-state-has-as-much-palestinian-blood-on-its-hands/.
95. "Torture Archipelago," *Human Rights Watch*. Last modified July 3, 2012. Accessed June 2, 2019. www.hrw.org/report/2012/07/03/torture-archipelago/arbitrary-arrests-torture-and-enforced-disappearances-syrias.
96. "The Jerusalem Brigade activists are not part of the Palestinian fabric and less than 13% of the number of Palestinian fighters," *Action Group for Palestinians of Syria*. Last modified August 18, 2016. Accessed June 2, 2019. www.actionpal.org.uk/ar/post/5536/ناشطون-لواء-القدس-ليس-من-النسيج-الفلسطيني-وأقل-من-14-من-عدد-مقاتلي-ه-فـلسطيني ين/م.جموعة-العمل-من-أجل-فـلسطينيي-سـورية.
97. Honest Bash, "The security role in the formation of al-Quds Brigade and the instrumentalization of the Palestinian question," *Maseer*. Last modified June 3, 2018. Accessed June 2, 2019. https://maseer.net/archives/1937.
98. Stephen Morton, "Sovereignty and necropolitics at the line of control," *Journal of Post Colonial Writing* 50, no. 1 (2014): 19–30.
99. David Kilcullen, Nate Rosenblatt, and Jwanah Qudsi, "Mapping the conflict in Aleppo, Syria," *Caerus*. Last modified February 2014. Accessed June 2, 2019. http://caerusassociates.com/wp-content/uploads/2014/02/Caerus_AleppoMappingProject_FinalReport_02-18-14.pdf, 17.
100. The Caerus researchers were able to count 1,462 checkpoints inside and around Aleppo. During that period, the regime controlled twice as many checkpoints as the opposition. See Kilcullen, Rosenblatt, and Qudsi, "Mapping the conflict in Aleppo, Syria," 19.
101. James Harkin, "Inside Aleppo, Syria's most war-torn city," *Newsweek*. Last modified August 19, 2015. Accessed June 2, 2019. www.newsweek.com/2015/08/28/syria-war-bombing-aleppo-364035.html.
102. Nahel Hariri, "The siege of Aleppo. Will the crossing open again?" *Al-Hayat*. Last modified November 4, 2015. Accessed June 2, 2019. www.alhayat.com/article/830427/حصار-حلب-الثاني-هل-يعود-المعبر.

103. Mohammed al-Khatieb, "Aleppans set up money transfer offices within the city," *Al-Monitor*. Last modified July 9, 2014. Accessed June 2, 2019. www.al-monitor.com/pulse/originals/2014/07/syria-aleppo-money-transfer-office.html.
104. "The boy of the crossing," *Ahewar*. Last modified December 16, 2013. Accessed June 2, 2019. www.ahewar.org/debat/show.art.asp?aid=391556&r=0.
105. Mayala, Hammam, and Salen, "Aleppo: People, place, and war."
106. Darwīsh and Abī Samrā, *ma'si halb* [*Tragedies of Aleppo*], 18–22.
107. Hussainiat are Shia places of worship.
108. "War of the marginal margins," *Shaam Network*. Last modified January 25, 2016. Accessed June 2, 2019. www.shaam.org/articles/studies-and-research/-ب-حرب-الهوامش-الحلبية.html.
109. Ibid.
110. Haroon Siddique and Brian Whitaker, "Syrian rebels 'overrun Aleppo police stations,'" *Guardian*. July 31, 2012. Accessed October 12, 2019. www.theguardian.com/world/2012/jul/31/syria-aleppo-fighting-goes-on-live.
111. John Caves, "Backgrounder: Syrian Kurds and the Democratic Union Party (PYD). Institute for the Study of War. December 6, 2012. Accessed October 17, 2019. www.understandingwar.org/sites/default/files/Backgrounder_SyrianKurds.pdf.
112. Luke Mogelson, "The river martyrs," *New Yorker*. April 29, 2013. www.newyorker.com/magazine/2013/04/29/the-river-martyrs.
113. Deen Sharp, "Urbicide and the arrangment of violence in Syria," in Deen Sharp and Claire Panetta (Eds.), *Beyond the Square: Urbanism and the Arab Springs* (New York: Terreform/Urban Research, 2016), 133.
114. Maksymilian Czuperski, Faysal Itani, Ben Nimmo, Eliot Higgins, Emma Beals, "Breaking Aleppo: hospital attacks," *Atlantic Council*, February 2017. Accessed October 12, 2019. www.publications.atlanticcouncil.org/breakingaleppo/hospital-attacks/.
115. Graham, Stephen, "Lessons in urbicide." *New Left review* 19(19): 63-77. January 2003.
116. Sawsan Abou Zainedin and Hani Fakhani, "Syria's urbicide: the built environment as a means to consolidate homogeneity," *The Aleppo Project*, July 2019. Accessed October 17, 2019. www.thealeppoproject.com/wp-content/uploads/2019/07/SyriasUrbicideSawsanAbouZainedinHaniFakhani2019.pdf.
117. Robert F. Worth, "Aleppo after the fall," *New York Times Magazine*, May 24, 2017. Accessed October 17, 2019. www.nytimes.com/2017/05/24/magazine/aleppo-after-the-fall.html.
118. Joseph Daher, "The paradox of Syria's reconstruction," *Carnegie Middle East Center*. September 4, 2019. Accessed October 12, 2019. https://carnegie-mec.org/2019/09/04/paradox-of-syria-s-reconstruction-pub-79773.

119. "Syria: defectors describe orders to shoot unarmed protesters," *Human Rights Watch*. Last modified July 9, 2011. Accessed June 2, 2019. www.hrw.org/news/2011/07/09/syria-defectors-describe-orders-shoot-unarmed-protesters#.
120. Mirjana Ristic, *Architecture, Urban Space and War: The Destruction and Reconstruction of Sarajevo* (London: Palgrave, 2018), 59.
121. "Snipers target mental patients in Syria," *News24*. Last modified October 10, 2013. Accessed June 2, 2019. www.news24.com/World/News/Snipers-target-mental-patients-in-Syria-20130110.
122. Ristic, *Architecture, Urban Space and War*, 53.
123. Ibid.
124. Zaher Sahloul, "The corridor of death," *Foreign Policy*. Last modified March 4, 2014. Accessed June 2, 2019. https://foreignpolicy.com/2014/03/04/the-corridor-of-death/.
125. Liqa Maki, "Beware the Sniper… You're in Aleppo," *Al-Jazeera*. Last modified September 5, 2013. Accessed June 2, 2019. www.aljazeera.net/news/reportsandinterviews/2013/5/9/حذر-من-القنص-فأنت-في-حلب.
126. Alanadol, "'Beware! Sniper': a sign of the smell of death in Bosnia and Aleppo," *El Watan News*. Last modified August 3, 2013. Accessed June 2, 2019. www.elwatannews.com/news/details/238126.
127. Lubna Salem, "A detailed protocol about how to avoid the sniper's eye," *Raseef22*. Last modified September 1, 2016. Accessed June 2, 2019. https://raseef22.com/article/72729-توكول-بروتصيلي-تتجنب-عين-آصالقن.
128. Olivier Laurent, "This is the surprising way some Syrians are protecting themselves from snipers," *Time*. Last modified March 27, 2015. Accessed June 2, 2019. http://time.com/3760455/aleppo-busses-syria/.
129. "Outwitting Syrian snipers," *Reuters*. Last modified February 4, 2013. Accessed June 2, 2019. www.reuters.com/news/picture/outwitting-syrian-snipers-idUSRTR3DCLA.
130. Sahloul, "The corridor of death."
131. "Snipers kill civilians at the death corridor," *Al Jazeera*. Last modified September 13, 2013. Accessed June 2, 2019. www.aljazeera.net/news/reportsandinterviews/2013/9/13/اصةالقن-يحصدون-المدنيين-بمعبر-الموت-بحلب.
132. Maki, "Beware the sniper."
133. Stephen Graham, *Vertical: The City from Satellites to Bunkers* (London: Verso, 2016).
134. Abigail Hauslohner and Ahmed Ramdan, "Ancient Syrian castles serve again as fighting positions," *Washington Post*. Last modified May 4, 2013. Accessed June 2, 2013. www.washingtonpost.com/world/middle_east/ancient-syrian-castles-serve-again-as-fighting-positions/2013/05/04/5d2bb176-b3f8-11e2-9a98-4be1688d7d84_story.html.
135. Graham, *Vertical*, 64.
136. Ibid.

CHAPTER 3

1. On the 100th anniversary of the Sykes–Picot Agreement, which divided the Levant into different countries after the defeat of the Ottoman Empire, ISIS released a statement in which it rejected the borders imposed by "Crusaders." For the full statement, see "The Islamic State and the world after Sykes Picot," *Internet Archive*. Undated._Accessed August 28, 2016. https://archive.org/stream/TheIslamicStateAndTheWorldAfterSykes Picot/The%20Islamic%20State%20and%20the%20World%20after%20Sykes-Picot_djvu.txt.
2. The Democratic Union Party (PYD), the left-wing Kurdish political party, aspires to control Rojava, a loosely defined territory in Northern Syria that has a large Kurdish population and a long border with Turkey. The Kurdish forces are trying to connect three separate cantons (Afrin, Kobane, and Jazira), while the Turkish military is trying to undermine their plans by occupying a strip of land that prevents the connection of the cantons. See "Charter of the Social Contract—self-rule in Rojava," *Peace in Kurdistan*. Last modified January 29, 2014. Accessed August 28, 2016. https://peaceinkurdistancampaign.com/charter-of-the-social-contract/.
3. The regime was facing challenging military, political, and economic obstacles that prevented it from regaining full control of the Syrian territory. In 2015, it suggested that its goal was to control what it labeled "Useful Syria," that is, the most densely populated and economically viable part of the country, which connects Damascus and Homs to the coast. The remaining territory was controlled by various groups, including the Kurds in the North, Islamists factions in Idlib, ISIS in Raqqa and Deir Ez-Zor, etc. See Hassan Mneimneh, "Will Assad create a "Useful Syria?" *Fikra Forum*, November 12, 2015. Accessed June 17, 2016. http://fikraforum.org/?p=8076#.V82DCWPZrds; and Robin Yassin-Kassab and Leila Al-Shami, *Burning Country: Syrians in Revolution and War* (London: Pluto Press, 2016), 145.
4. Salwa Ismail, "The Syrian uprising: imagining and performing the nation," *Studies in Ethnicity and Nationalism* 11 (2011): 538–49; and Manal al-Natour, "Nation, gender, and identity: children in the Syrian Revolution 2011," *Journal of International Women's Studies* 14, no. 5 (2013): 28–49.
5. The distinction between official and popular nationalisms was first suggested by Hugh Seton-Watson in *Nations and States: An Inquiry into the Origins of Nations and the Politics of Nationalism* (Boulder, CO: Westview Press, 1977) and then popularized by Benedict Anderson in his seminal work, *Imagined Communities: Reflections on the Origin and Spread of Nationalism* (New York: Verso, 1983). Anderson explains that official nationalism was originally a conservative and top-down ideology in reaction to the spontaneous popular nationalism that was thriving in Europe in the early nineteenth century—that official nationalism is the "willed merger of national and dynastic empire—[…] developed *after*, and

in reaction to, the popular national movements proliferating in Europe since the 1820s"; Anderson, *Imagined Communities*, 86.
6. Frantz Fanon, *The Wretched of the Earth*, trans. Constance Farrington (New York: Grove Press, 1963), 123–4.
7. Anna M. Agathangelou, "The living and being of the streets: Fanon and the Arab uprisings," *Globalizations* 9, no. 3 (2012): 451–66.
8. Hamid Dabashi, *The Arab Spring: The End of Postcolonialism* (London: Zed Books, 2012).
9. See Fanon's "The pitfalls of national consciousness," in *The Wretched of the Earth*, trans. Constance Farrington.
10. Nigel C. Gibson, *Fanon: The Postcolonial Imagination* (London: Polity, 2003), 184.
11. Frantz Fanon, *The Wretched of the Earth*, trans. Richard Philcox (New York: Grove Press, 2004), 237–8.
12. See Philip S. Khoury, *Syria and the French Mandate: The Politics of Arab Nationalism, 1920–1945* (Princeton, NJ: Princeton University Press, 2014); Elizabeth Thompson, *Colonial Citizens* (New York: Colombia University Press, 2000); Michael Provence, *The Great Syrian Revolt and the Rise of Arab Nationalism* (Austin, TX: University of Texas Press, 2005); and James L. Gelvin, *Divided Loyalties: Nationalism and Mass Politics in Syria at the Close of Empire* (Berkeley, CA: University of California Press, 1998).
13. Fanon, *The Wretched of the Earth*, trans. Richard Philcox, i.
14. Neil Lazarus, *Nationalism and Cultural Practice in the Postcolonial World* (Cambridge: Cambridge University Press, 2003), 78–103.
15. Ibid., 86.
16. Philip S. Khoury, *Urban Notables and Arab Nationalism: The Politics of Damascus, 1860–1920* (Cambridge: Cambridge University Press, 1983), 67.
17. Roger Owen, "Arab Nationalism Arab Unity and Arab Solidarity," in T. Asad and R. Owen (Eds.), *Sociology of the "Developing Societies": The Middle East* (New York: Monthly Review Press, 1983), 16–22.
18. Provence, *The Great Syrian Revolt*; and Gelvin, *Divided Loyalties*.
19. Gelvin, *Divided Loyalties*.
20. *Qabadyat* is plural for *qabaday*, the local strongmen who have a good reputation in the neighborhood and are willing to protect their neighbors.
21. James L. Gelvin, "The other Arab nationalism: Syrian/Arab populism in its historical and international context," in G. Israel and J. James (Eds.), *Rethinking Nationalism in the Arab Middle East* (New York: Columbia University Press, 1999), 231–48.
22. Fanon, *The Wretched of the Earth*, trans. Richard Philcox, 144.
23. Provence, *The Great Syrian Revolt*.
24. Both Gelvin, *Divided Loyalties* and Provence, *The Great Syrian Revolt* suggest that Syrian nationalism should be understood as a non-essentialist discourse with unstable and historically fluid meanings.

25. Harika means fire in Arabic. The French bombed the neighborhood and burnt it down on October 18, 1925 to punish the population because of its support for the revolt. Approximately 1,500 residents were killed.
26. Provence, *The Great Syrian Revolt*, 17.
27. Khoury, *Syria and the French Mandate*, 237.
28. Gibson, *Fanon*, 184.
29. Ibid., 182.
30. For a history of post-independence nationalism, see Youssef Chaitani, *Post-Colonial Syria and Lebanon: The Decline of Arab Nationalism and the Triumph of the State* (New York: Palgrave Macmillan, 2007); and Adeed Dawisha, *Arab Nationalism in the Twentieth Century: From Triumph to Despair* (Princeton, NJ: Princeton University Press, 2003).
31. The Syrian regime's official discourse is one that celebrates Arab nationalism and Arab unity. This discourse is utilized pragmatically by claiming non-sectarian and inclusion. It pretends to be impartial, and does not differentiate between citizens from the various religions and sects. However, the discourse is actually ambivalent, since all the strategic positions of the state are controlled by members of the Alawi sect—an ambivalent discourse that is both inclusive and anti-sectarian, and is strategically deployed against what it wants the minorities to perceive as the fundamentalism and extremism of the Sunni majority. It also insists that only the Syrian branch of the Baath party, which is controlled by the Syrian president, is in a position to define and police the meaning of Arab nationalism.
32. Shayna Silverstein shows that Syrian protesters re-adapted the traditional Dabke dance, what she terms as a "radical Dabke," to "take back the streets, as well as the cultural symbols of their national heritage," see Silverstein, "Syria's Radical Dabke," *Middle East Report* 263 (2012), 38. Tahira Yaqoob explains that many Syrian artists use their art as a weapon of war, and as such create a new language and community, see Tahira Yaqoob, "Syrian artists using their medium as a weapon," *The National*, March 7, 2013. Accessed June 24, 2016. www.thenational.ae/arts-culture/art/the-syrian-artists-using-their-medium-as-a-weapon#full2013.
33. Fanon, *The Wretched of the Earth*, trans. Constance Farrington, 204.
34. Misbar Syria, "Demonstration of Syrian anger in Damascus 17 February 2011," (Arabic). *YouTube*. Video File. February 17, 2001. Accessed May 13, 2019. www.youtube.com/watch?v=qDHLsU-ik_Y.
35. Almanbegee, "Demonstration of the city of Manbij on Good Friday," (Arabic). *YouTube*. Video File. April 23, 2011. Accessed May 13, 2019. www.youtube.com/watch?v=mukC1OkE9oc&frags=pl%2Cwn.
36. Javier Espinosa, "Syria: defiance of village where army killed 39 from a single family," *Guardian*. Last modified June 8, 2012. Accessed May 13, 2019. www.theguardian.com/world/2012/jun/08/ syria-defiance-village-taftanaz-massacre.
37. Katie Paul, "Syrian rebels, regime offer dueling tales of Karm al-Zeitoun Massacre," *The Daily Beast*. Last modified March 13, 2012. Accessed May 13,

2019. www.thedailybeast.com/syrian-rebels-regime-offer-dueling-tales-of-karm-al-zeitoun-massacre?ref=scroll.
38. Almanbegee, "Night demonstration in the city of Manbij 14/6/2011," (Arabic). *YouTube*. Video File. June 15, 2011. Accessed May 14, 2019. www.youtube.com/watch?v=Vsa6frboFV4.
39. Manbej1, "Attempt to break the statue of the tomb of Hafez al-Assad in Manbij 26/11/20," (Arabic). *YouTube*. Video File. November 28, 2011. Accessed May 14, 2019. www.youtube.com/watch?v=QepBUA1TfPo&frags=pl%2Cwn.
40. Saleh Al-Dandn, "Aleppo Manbij: Raising the flag of Independence at the entrance to the city," (Arabic). *YouTube*. Video File. April 9, 2012. Accessed May 14, 2019. www.youtube.com/watch?v=MUfCWLy9QJo.
41. Many interviewees reiterated this point. They believe that the revolt would not reach Manbij because of the supposed solid loyalist base present in the city.
42. Saleh Al-Dandn, "Manbij: a night demonstration in Al-Karama neighborhood Nasra al-Hula 26-5-2012," (Arabic). *YouTube*. Video File. May 26, 2012. Accessed May 14, 2019. www.youtube.com/watch?v=FmbdaKW4gqg&frags=pl%2Cwn.
43. This dynamic is evidently not unique to Manbij; many cities had a similar trajectory.
44. Fanon, *The Wretched of the Earth*, trans. Constance Farrington, 168.
45. Agathangelou, "The living and being of the streets," 463.
46. Martin Luther King, Jr. "Letter from Birmingham Jail," April 16, 1963. Accessed June 15, 2016. www.africa.upenn.edu/Articles_Gen/Letter_Birmingham.html.
47. Frantz Fanon, *Black Skin, White Masks*, trans. Charles Lam Markmann (New York: Grove Press, 1967), 8.
48. John Holloway, "Dignity's Revolt," *libcom.org*. Last modified November 1, 2005. Accessed June 22, 2016. https://libcom.org/library/dignitys-revolt-john-holloway.
49. The Islamic Brigade in Aleppo and its countryside, "Change the Circle of Abu Pearl in the city of Manbij to Roundabout Omar ibn al-Khattab," (Arabic). *YouTube*. Video File. April 22, 2013. Accessed May 14, 2019. www.youtube.com/watch?v=GMGp5yvqv1k.
50. Kelly McEvers, "Slain Syrian filmmaker traded study for 'revolution,'" *NPR: Obituaries*. Last Modified May 29, 2012. Accessed May 14, 2019. www.npr.org/2012/05/29/153937342/student-helped-the-world-see-inside-a-ravaged-syria.
51. Scott Peterson, "Syrian activists galvanized by killing of Kurdish leader," *Christian Science Monitor*. Last modified October 11, 2011. Accessed May 14, 2019. www.csmonitor.com/World/Middle-East/2011/1011/Syrian-activists-galvanized-by-killing-of-Kurdish-leader.
52. Hugh Macleod and Annasofie Flamand, "Tortured and killed: Hamza al-Khateeb, age 13," *Al Jazeera*. Last modified October 11, 2011. Accessed

May 14, 2019. www.aljazeera.com/indepth/features /2011/05/20115318 5927813389.html.

53. DEMOCRACY 4 SY. "Aleppo Youth Institute for Syria," (Arabic). *YouTube*. Video File. December 6, 2015. Accessed May 14, 2019. www.youtube.com/watch?v=IXx4BMHBLTU&frags=pl%2Cwn.

54. Razan Ghazzawi, "Seeing the women in revolutionary Syria," *Open Democracy*. Last modified April 8, 2014. Accessed May 14, 2019. www.opendemocracy.net/north-africa-west-asia/razan-ghazzawi/seeing-women-in-revolutionary-syria; Carol Morello, "Role of Syrian women evolves as war rages on," *Washington Post*. Last modified January 10, 2013. Accessed May 14, 2019. www.washingtonpost.com/world/middle_east/role-of-syrian-women-evolves-as-war-rages-on/2013/01/09/5308512e-559b-11e2-bf3e-76c0a789346f_story.html?utm_term=.d49ea712dc7c; and Layla Saleh, "In Syria, from fighting to blogging, the many roles of women," *The Conversation*. Last modified October 6, 2016. Accessed May 15, 2019. https://theconversation.com/in-syria-from-fighting-to-blogging-the-many-roles-of-women-65176.

55. DEMOCRACY 4 SY. "Aleppo Youth Village, 12-12-2013: Future Youth Authority National Patient Hospital," (Arabic). *YouTube*. Video File. October 13, 2013. Accessed May 14, 2019. www.youtube.com/watch?v=CGYyuuGnbLo.

56. Muhamad Alealli, "Youth Futures authority: footprint of volunteer work in the city of Manbij," *OrientNet*. Last modified November 22, 2013. Accessed May 15, 2019. www.orient-news.net/ar/news_show/6335.

57. DEMOCRACY 4 SY, "Aleppo Future Youth for Syria," (Arabic). *YouTube*. Video File. October 30, 2012. Accessed May 15, 2019. www.youtube.com/watch?v=o2_M8_tOYuI&t=132s&frags=pl%2Cwn.

58. Saleh Al-Dandn, "Aleppo–Manbij, 13-9-2012: Future Youth for Syria," (Arabic). *YouTube*. Video File. September 13, 2012. Accessed May 15, 2019. www.youtube.com/watch?v=cUjzPfeaoag&frags= pl%2Cwn.

59. Moetaz AboRyad, "Aleppo countryside: within the performance of honoring the martyrs, a free play by a young Syrian 20/9/2013," (Arabic). *YouTube*. Video File. September 20, 2013. Accessed May 15, 2019. www.youtube.com/watch?v=VdgwWFfLK7c.

60. "Syrian Street Festival 2013," *Ashar*. Accessed May 15, 2019. http://ashar3.com/ar/portfolio/ةيلافتحإ-عراشلأ-يروسلا-2013/.

61. Borzou Daraghi, "Syrian rebels raise a flag from the past," *Financial Times*, December 30, 2011. Accessed June 25, 2016. www.ft.com/cms/s/0/6c332676-32f4-11e1-8e0d-00144feabdco.html#axzz4JPBrIeLI.

62. Alser ax, "Demonstration against the State of Iraq and the Islamic Shami in the city of Mnabh Brive Aleppo," (Arabic). Video file. *YouTube*. July 9, 2013. Accessed June 7, 2016. www.youtube.com/watch?v=OcAN--iVES o&spfreload=10.

63. Abu Bakr al-Baghdadi explains in his Ramadan sermon in 2014:

> Muslims today have a loud, thundering statement, and possess heavy boots. They have a statement that will cause the world to hear and

understand the meaning of terrorism, and boots that will trample the idol of nationalism, destroy the idol of democracy and uncover its deviant nature...

"Foreword," *Dabiq* 12 (2015): 3, https://azelin.files.wordpress.com/2015/11/the-islamic-state-e2809cdc48.

64. Interviewee. Interview conducted in Manbij, Syria. December 27, 2013.
65. "FSA 'closing in' on Manbij as Islamic State in retreat," *Syria Direct*. Last modified February 15, 2015. Accessed May 15, 2019. https://syriadirect.org/news/fsa-closing-in-on-manbij-as-islamic-state-in-retreat/.

CHAPTER 4

1. This chapter is based on ethnographic research and interviews conducted in Manbij and other cities in Northern Syria during several extended trips in 2011–2014.
2. Elizabeth Thompson, "The climax and crisis of the colonial welfare state in Syria and Lebanon during World War II," in Steven Heydemann (Ed.), *War, Institutions, and Social Change in the Middle East* (Berkeley, CA: University of California Press, 2000), 75–6.
3. Ibid.
4. Joel Beinin, *Workers and Peasants in the Modern Middle East* (Cambridge: Cambridge University Press, 2001), 121.
5. Steven Heydemann, *Authoritarianism in Syria: Institutions and Social Conflict, 1946–1970* (Ithaca, NY: Cornell University Press, 1999), 47.
6. Ross Burns, *Aleppo: A History* (New York: Routledge, 2017), 283.
7. Majid Khadduri, "Constitutional development in Syria: with emphasis on the Constitution of 1950," *Middle East Journal* 5, no. 2 (Spring 1951): 137–60.
8. Heydemann, *Authoritarianism in Syria*, 48–50.
9. Ibid., 32.
10. Ibid., 51–2.
11. Hanna Batatu, *Syria's Peasantry, the Descendants of Its Lesser Rural Notables, and Their Politics* (Princeton, NJ: Princeton University Press, 1999), 44.
12. Heydemann, *Authoritarianism in Syria*, 111–12.
13. Ibid., 112.
14. Ibid., 112–17.
15. Ibid., 193.
16. Raymond Hinnebusch, "The Ba'ath's agrarian revolution (1963–2000)," in *Agriculture and Reform in Syria* (Boulder, CO: Lynne Rienner Publishers, distributed for the University of St. Andrews Center for Syrian Studies, 2011).
17. Beinin, *Workers and Peasants in the Modern Middle East*, 133.
18. Batatu, *Syria's Peasantry*, 47.
19. Thomas Nail, "What is an assemblage?" *SubStance* 46, no. 1 (2017), 24.

20. Ibid.
21. Deleuze and Guattari identify four types of assemblages: territorial, state, capitalist, and nomadic. Each one of these assemblages emerges at a different historic conjuncture and has its own specificities.
22. Gilles Deleuze and Felix Guattari, *A Thousand Plateaus: Capitalism and Schizophrenia* (Minneapolis, MN: University of Minnesota Press, 1987), 478.
23. Massimiliano Trentin, "Modernization as state building: the two Germanies in Syria, 1963–1972," *Diplomatic History* 33, no. 3 (June 2009), 492.
24. Ibid., 496.
25. Ibid., 494.
26. Heydemann, *Authoritarianism in Syria*, 195–6.
27. Myriam Ababsa, "Fifty years of state land distribution in the Syrian Jazira: agrarian reform, agrarian counter-reform and the Arab Belt Policy (1958–2008)," in Habib Ayeb and Reem Saad (Eds.), *Agrarian Transformation in the Arab World: Persistent and Emerging Challenges* (Cairo: American University in Cairo Press, 2009), 36–7.
28. Ibid.
29. Ibid.
30. Ibid., 39.
31. Batatu, *Syria's Peasantry*, 122.
32. Heydemann, *Authoritarianism in Syria*, 202.
33. Ababsa, "Fifty years of state land distribution in the Syrian Jazira," 37.
34. Trentin, "Modernization as state building," 495.
35. Batatu, *Syria's Peasantry*, 37.
36. Ibid., 51.
37. Ibid.
38. Nabil Sukkar, "The crisis of 1986 and Syria's plan for reform," in Eberhard Kienlp (Ed.), *Contemporary Syria: Liberalization between Cold War and Cold Peace* (London: British Academic Press, 1994), 23.
39. Hinnebusch, "The Ba'ath's agrarian revolution (1963–2000)," 4.
40. Ibid., 5.
41. Batatu, *Syria's Peasantry*, 92.
42. Hinnebusch, "The Ba'ath's agrarian revolution (1963–2000)," 3.
43. Deleuze and Guattari, *A Thousand Plateaus*, 453.
44. "Agriculture: Syrian food basket from retreat to disaster," *Harmoon*. September 2017. Accessed July 4, 2019. https://harmoon.org/ةعارزلا-ةلس-الاتراجع-من-السورية-الغذاء/.
45. "Crop and Food Security Assessment Mission to the Syrian Arab Republic," *FAO/WFP*. July 5, 2013. Accessed June 2, 2019, 9. www.fao.org/docrep/018/aq113e/aq113e.pdf.
46. Ibid., 21.
47. "Syrian Civil War cut wheat harvest to its worst level," *World Bulletin*, July 25, 2013. Accessed June 1, 2019. www.worldbulletin.net/news/113852/syrian-civil-war-cut-wheat-harvest-to-its-worst-level.

48. "Syrian Arab Republic wheat production by year," *Index Mundi*. Accessed June 5, 2019. www.indexmundi.com/agriculture/?country=sy&commodity =wheat&graph=production.
49. Ibid.
50. Ibid.
51. Jessica Barnes, "Managing the waters of Ba'th country: the politics of water scarcity in Syria," *Geopolitics* 14, no. 3 (2009), 524.
52. Ibid., 525.
53. Food and Agriculture Organization of the United Nations (FAO), "Syrian agriculture at the crossroads," (2003), 101. Accessed June 1, 2019. www. napcsyr.gov.sy/dwnld-files/fao_publications/sac/syrian_agriculture_at_ the_cross_roads_en.pdf.
54. Aden Aw-Hassana, Fadel Ridab, Roberto Telleria, and Adriana Bruggeman, "The impact of food and agricultural policies on groundwater use in Syria," *Journal of Hydrology* 513, (May 26, 2014): 204–15.
55. "The General Company for Mills." Accessed July 18, 2019. http://mills.gov.sy.
56. Maha El Dahan and Jonathan Saul, "Syria taps world wheat market as stocks run down," *Reuters*. March 30, 2015. Accessed June 7, 2019. www. reuters.com/article/2015/03/30/syria-food-imports-idUSL5NoW81 PQ20150330.
57. FAO, "Syrian agriculture at the crossroads," 41.
58. Ibid., 42.
59. Ibid., 48.
60. See Lisa Wedeen, "Acting 'as if': the story of M," in *Ambiguities of Domination: Politics, Rhetoric, and Symbols in Contemporary Syria*, 1st edn (Chicago, IL: University of Chicago Press, 1999).
61. Adnan Abdul Razak, "Syria's wheat crop is the worst in 40 years," *Al-Araby*. June 24, 2013. Accessed June 28, 2019. www.alaraby.co.uk/economy/ عام-40-منذ-الأسوأ-هو-سورية-في-القمح-محصول/24/6/2014.
62. "Bread shortages rising," *The New Humanitarian*. December 13, 2012. Accessed June 3, 2019. www.thenewhumanitarian.org/analysis/2012/12/13/ bread-shortages-rising.
63. "Crop and Food Security Assessment Mission to the Syrian Arab Republic," https://reliefweb.int/report/syrian-arab-republic/special-report-faowfp-crop-and-food-security-assessment-mission-syrian-3.
64. Sam Heller, "How Assad is using sieges and hunger to grab more of the 'Useful Syria,'" *Vice*. January 14, 2016. Accessed May 2, 2019. www.vice. com/en_us/article/7xadz9/how-assad-is-using-sieges-and-hunger-to-grab-more-of-the-useful-syria.
65. Mohammad Kanfash and Ali al-Jasem, "Syrians are watching their crops burn. These crimes of starvation must end," *Guardian*. July 11, 2019. Accessed July 27, 2019. www.theguardian.com/global-development/2019/ jul/11/syrians-are-watching-their-crops-burn-these-starvation-crimes-must-end.

66. Gianluca Serra, "Over-grazing and desertification in the Syrian steppe are the root causes of war," *The Ecologist*. June 5, 2015. Accessed July 2, 2019. www.theecologist.org/News/news_analysis/2871076/overgrazing_and_desertification_in_the_syrian_steppe_are_the_root_causes_of_war.html.
67. Jan Selbya, Omar S. Dahi, Christiane Fröhlich, and Mike Hulmee, "Climate change and the Syrian Civil War revisited," *Political Geography* 60 (September 2017): 232–44. https://doi.org/10.1016/j.polgeo.2017.05.007.
68. Nail, "What is an Assemblage?", 33.
69. "Syria: government attacking bread lines," *Human Rights Watch*. August 30, 2012. Accessed June 29, 2019. www.hrw.org/news/2012/08/30/syria-government-attacking-bread-lines.
70. Ibid.
71. "Syria sees money in bumper harvest, but getting to it is hard," *Reuters*. May 20, 2015. Accessed June 3, 2019. www.reuters.com/article/2015/05/20/mideast-crisis-syria-wheat-idUSL5N0YA3VB20150520.
72. Ibid.

CHAPTER 5

1. This chapter was initially published in Thanassis Cambanis and Michael Wahid Hanna (Eds.), *Arab Politics Beyond the Uprisings: Experiments in an Era of Resurgent Authoritarianism* (Washington, DC: Brookings Institution Press, 2017).
2. "Manbij Perishes Silently" (Arabic), *Syria Untold*, July 20, 2016.
3. As I have noted, I am referring to the Manbij activists who were instrumental in setting up the RC as "revolutionaries," using the term they used to describe themselves. Similarly, I use other concepts popular among the activists throughout this chapter, such as "liberation." Although I acknowledge the risk of using normative terms such as these to describe events over which there is still a great deal of disagreement even within Syria, it would be a poor representation of my research to use excessively impartial vocabulary.
4. I borrow Haitian intellectual Michel-Rolph Trouillot's terminology for his country's "unthinkable revolution" at the turn of the nineteenth century, which was ignored in much of the world due to racism. Analogously, Western thought has ignored the accomplishments of Syrian revolutionaries and—in its ignoring of them—telegraphed the belief that a revolutionary process is inconceivable in an Arab country such as Syria.
5. The FSA was very marginal in Manbij and did not play any notable role until after the liberation of the city. See Christoph Reuter and Abd al-Kadher Adhun, "Rebels make a go of governing in liberated city," *Der Spiegel*, October 2, 2012.
6. In general, the FSA in Manbij was ill equipped and underfunded; it had less than 20 machine guns on the day the residents expelled the police and security from the city.

7. For more on Egypt's local popular committees, see Cilja Harders and Dina Wahba, "New neighborhood power: informal popular committees and changing local governance in Egypt," *The Century Foundation*. February 14, 2017. Accessed June 2, 2019. https://tcf.org/content/report/new-neighborhood-power/?agreed=1; and Aysa El-Meehy, "Governance from below: comparing local experiments in Egypt and Syria after the uprisings," *The Century Foundation*. February 7, 2017. Accessed June 2, 2019. https://tcf.org/content/report/governance-from-below/.
8. For more information, see Haian Dukhan, "Tribes and tribalism in the Syrian uprising," *Syria Studies Journal* 6, no. 2 (2014): 1–28.
9. Amjad Altinawi, "The security establishment executes 'The Prince,'" (Arabic), *All4Syria*, April 21, 2015.
10. Zana Miso, "Faruq Brigades leave Homs under shelling and head to al-Hasakeh to Fight Kurdish Forces," (Arabic), *YouTube*, July 23, 2013.
11. Rania Abouzeid, "Syria's up-and-coming rebels: who are the Farouq Brigades?," *Time*, October 5, 2012.
12. Manbij and other regions in northern Aleppo province were dominated by the Naqshbandiyya Sufi order. Salafism had a contentious relationship with Sufism in Syria. For an in-depth discussion, see the important work of Itzchak Weismann on Salafism and Sufism in Syria.
13. Christina Zdanowicz, "Aleppo is being destroyed by the silence of the world," *CNN*, December 13, 2016.
14. Idrees Ahmad, "Russia Today and the post-truth virus," *Pulsemedia*, December 15, 2016.
15. Ibid.
16. Kim Sengupta, "Turkey and Saudi Arabia alarm the West by backing Islamist extremists the Americans had bombed in Syria," *The Independent*, May 11, 2015.
17. Even outlets opposed to Assad and friendly to the opposition tend to overemphasize the geopolitical aspect of the revolution. See the following articles from major outlets that have been generally critical of the Assad regime: As'ad Abukhalil, "The Left and the Syria debate," *Jadaliyya*, December 10, 2016; Max Fisher, "In Syrian War, Russia has yet to fulfill superpower ambitions," *New York Times*, September 24, 2016; Nafeez Ahmed, "Syria intervention plan fueled by oil interests, not chemical weapon concern," *Guardian*, August 30, 2013; David Oualaalou, "Russia outsmarts the U.S. in Syria: a new geopolitical outlook," *Huffington Post*, October 13, 2016.

Index

Aflaq, Michel, 11, 12, 21, 97, 105
Agamben, Giorgio, 10, 16, 17–20, 24, 26–7, 32, 36, 41
  bare life, 18, 20–1, 24; bios, 20, 24, 41; homo sacer, 33; state of exception, 10, 17–20, 24, 26, 36, 38; zoe, 20, 24, 32, 41
agrarian reform, 11, 120, 121–32 see agriculture; also see Akram Hourani; counter, 59–60, 135, 142; land reform, 12, 44, 51, 121–32, 134, 135, 142 see social classes
agriculture
  cotton, 132, 134; droughts, 2, 52, 60, 134, 137; hydrological projects, 121, 128, 131, 133, 135–6, 138; irrigation, 2, 121, 128, 129, 132–6, 138, 142; over-exploitation, 137; see social classes; wells, 134, 135, 137; wheat, 9, 120, 132–7, 139, 141–3
Ahrar al-Sham, 116–7, 119, 140–1, 149, 151, 153, 155
Alawites, 13, 21, 31, 49, 50, 56, 57, 102
Aleppo, 43–94
  Ashrafieh (district), 64, 65; Armenians, 58–9; Bab el-Nayrab (district), 45; al-Berri clan, 63, 70, 72; Castello road, 65; checkpoints, 55, 58, 64, 68–71, 72; Christians, 58; the Citadel, 57, 78; demographics, 53–68; Free Students Union, 53; Free Syrian Army (FSA), 53, 57, 59, 60–1, 63–4, 65, 67, 72, 76, 78; French mandate, 46–8; al-Hamdanieh (district), 56; Handarat camp, 66; Hannano barracks, 72; informal settlements, 51, 53, 59–60, 64, 79; medical facilities, 32, 35, 53, 62, 73; municipal building, 69, 76, 77; Muslim Brotherhood, 49–52; nationalism, 46; al-Nayrab camp, 66; notables, 46, 48–9, 79; notable politics, 45; Old City, 46, 48–9; 51, 53, 71, 78; Ottoman, 44–5; Queiq river, 36, 72; Palestinians, 66–7; rebellion (1980–1), 49–52; riot (1850), 45; post-independence, 48–9; Saadallah al-Jabiri square, 52, 53, 71, 77; Sakhour (district), 53, 61; sectarianism, 56–9; al-Shaar (district), 57; Shabiha, 53, 61–2, 63, 70–1; Sheikh Maqsoud (district), 64–6, 72; siege, 61, 65–6, 69; Suleimaniya (district), 58, 59; Tansiqiat (coordination committees), 53; Tanzimat, 44–5; tribes, 62–4; United Judicial Council (UJC), 57; university, 53, 58, 60–2, 71, 72; urbanism, 47–8; volcano of Aleppo, 52; youth, 60–2
al-Ali, Shaykh Salih (Alawi leader in 1919), 102
al-Arsuzi, Zaki (Arab nationalist), 97
al-Assad, Bashar, 23, 29, 31, 52, 56, 59, 73, 75, 115–6, 134, 145, 146
al-Assad, Hafez, 10, 12–14, 15, 16, 21–2, 38–9, 59, 62–3, 66, 67, 70, 116, 119, 127, 130–2, 134
al-Asaad, Khaled (Syrian archeologist), 3
al-Atassi, Nureddin, 130
al-Atrash, Sultan, 105
Attar, Essam, 22
Azaz, 154
al-Azmeh, Yousef, 14

Baath party, see political parties

Bab al-Nayrab, *see* Aleppo
al-Balous, Sheikh Wahid, 31
Banshoya, Gyoji (Japanese urban planner), 49
barrel bomb, 8, 33, 34, 35, 72–3, 76, 116
Bayat, Asef, 5–6
Benjamin, Walter, 17, 18–9
  revolutionary violence, 19
Berri clan, 30, 63, 70, 72
  see Shabiha
al-Bitar, Salah el-Din, 14, 97, 105
Bookchin, Murray, 64
al-Bou Banna, *see* Manbij
bread, 120–43
  see agriculture; bakeries, 35, 72, 116, 135, 138, 139–40, 142, 152; brigades, 140; crisis, 135–6; distribution, 8; economy of, 120, 132–4, 138; General Company for Baking (GCB), 134–5; General Company for Mills (GCM), 134–5, geography of, 9, 140–2; grain mills, 116, 134–5, 138, 139–41, 149, 152; lethal weapon, 120, 141–2; lines, 120, 139; Manbij, 137–42; *see also* politics; riots, 9; price, 121, 132, 135, 136, 138–9, 150; strategic commodity, 116; supply, 135, 138–9; Syria's food basket, 136
Busha'ban tribe, *see* tribes

Césaire, Aimé, 27
checkpoint, 8, 28–9, 34, 41, 141
  Aleppo, 55, 58, 64, 68–71, 72; Bustan al-Qasr (District), 76; Karaj al-Hajez (death crossing), 29, 69
Circassians, 103
committees, 101, 108–9
  Clandestine, 108–9; Neighborhood, 101, Popular, 101, 147; tansiqiat (local), 53,

Damascus, 3, 11, 14, 29, 33, 36, 52, 66–7, 101–3, 108

Damascus Spring, 22–3
Danger Brothers (French urban planners), 46–7
Dara'a, 12, 61, 74, 109, 114
Daraya (Damascus suburb), 36
Deir ez-Zor, 12, 133
Deleuze, Gilles, 6, 126–8, 132
  deterritorialization, 126–9, 134, 138; philosophy of immanence, 126; reterritorialization, 127–8, 130, 138; assemblage, 120–1, 126–9, 132, 134, 138–43
Democratic Arab Socialist Union, 22
Democratic Union Party (PYD), *see* political parties
Druze, 31, 102–3
  Balous, Sheikh Wahid, 31; Jabal al-, 14, 102; officers, 21; revolt, 63, 75, 102; Sheikhs of dignity, 31; Suweida, 31,

economy
  bread, 132–4; crisis, 73, 131; decree no.10, 131; developmentalism, 106, 124–5, 128, 131; liberalization, 51, 52, 59, 131, 134–5; policies, 9, 50, 52, 130; privatization, 131; reform, 125; state protectionism, 137; taxation, 44, 121, 149
Eurocentrism, 4, 6, 20, 27, 145

Fanon, Frantz, 96–9, 101–2, 106, 109, 111, 113, 114
  see new humanism; zones of nonbeing, 113–4
Fayad, Shafiq (general in the Syrian army), 51
Faysal, King, 46, 101
Free Syrian Army (FSA), 7, 30, 53, 57, 59, 60, 61, 63–4, 65, 67, 71, 72, 76, 78, 109, 117, 140, 141, 146–7, 149, 151, 152, 153
  Farouq Brigades, 149; Jaysh al-Islam, 155; Revolutionaries of Manbij battalion, 151

INDEX

French mandate, 37, 38, 46–8, 102–5, 121
Colonial violence, 27, 62; colonial urbanism, 46–9, 78–9

General Peasants Union (GPU), 130
geopolitics, 4–5, 143
Great Syrian Revolt, see revolt
Guattari, Félix, 126, 127, 128, 132.
Gutton, André (French urban planner), 48–9, 51

Hadid, Marwan, 50
Hamid, Sultan Abdul, 100
Hannano, Ibrahim, 102
al-Hariri, Rafic, 14
al-Hasakah, 61, 133.
Hatoum, Salim (Druze Major), 21
al-Hourani, Akram, 12, 21, 121–2, 124
Hussein, Sharif, 100

Idlib, 30, 67, 111
Islam Army (Jaysh al-Islam), 155
Islamic State in Iraq and Syria (ISIS), 2, 3–4, 36, 56, 65, 95, 107, 108, 116–9, 144–5, 150, 151–4

Jadid, Salah, 12, 13, 21, 130
Jaysh al-Islam, see Islam Army
Jumblat, Kamal, 14

Kaileh, Salameh (Palestinian-Syrian intellectual), 1
Khalifeh, Mustafa (novelist and former political prisoner), 37, 39
al-Khatib, Hamza (adolescent killed by the regime), 115
Kurds
Arab belt, 59, 64, 65; Democratic Union Party (PYD), 31, 59, 64–5, 154; Kurdistan Workers' Party (PKK), 64; rebellion, 63, 65, 75; see Tammo, Mashaal

Latakia, 12, 31, 67

Le Corbusier (Swiss/French urban planner), 78–9
legal system,
Arab Unified Penal Code, 153; constitution (1973), 15, 21, 153; constitution (2012), 15, 22, 50, 122; counter-terrorism laws, 15, 24; counterterrorism court, 24; customary law, 62, 153, 154; emergency law, 14, 15; field court, 15; international law, 16, 20; law no.49, 16; law no.143, 124–5; martial law, 14; Revolutionary court, 152–4; Sharia court, 57; Supreme State Security Court (SSSC), 15; Syrian penal code, 23, 152, 153, 154

Ma'art al-Nouman, 29
macropolitics, 4, 7, 30
Manbij, 7, 9
Ahrar al-Sham, 116; battalion, 151; al-Bishir, Mohammed (local leader), 147–8; Bou Banna clan, 148; Council of the Trustees of the Revolution, 150–2; al-Dibo, Sheikh Said Mohammed, 152; Free Syrian Army (FSA), 117; Future youth, 115; Home (cultural center), 118–9; Islamic State, 116–9; Jaysh al-Islam, 155; My City is My Home (campaign), 115, 118; nationalism, 106–19; al-Nusra, 116; politics of bread, 137–43; micropolitics, 144–56; the Prince (military commander), 149; revolutionary council, 146–50; revolutionary court, 152–4; see also legal system; Sharia Council, 154; Sufi, 152; women, 115
Mando, Jalal, 33
Maydan Quarter, 102, 105
Maysaloun (battle of), 14
Mbembe, Achille, 10, 24–7, 32, 35
micropolitics, 4, 5, 7–8, 9, 30–1, 138, 144–56, 157

Mukhabrat, *see* security branches

Nakkba, 123
Nasser, Abdel, 11–12, 14, 106, 120, 124–5
National Democratic Rally, 22–3
Nationalism, 6, 46, 95–119
    Anti-colonial, 97, 99, 102; Arab, 95, 103; bourgeois, 97; New humanism, 97–9, 102; official/state, 96, 97, 99–100, 104–5; Pan-Arabism, 11, 97, 100, 106; popular, 95, 97, 99–100, 101–4; Syrian, 97, 99
necropolitics, 8, 24–8, 30, 31, 32–3, 34, 35, 37–8, 49, 40
neoliberalism, 73, 77, 97, 115
New Social Movement (theory), 5–6
Nietzsche, Friedrich, 6
al-Nusra Front (or Jabhat al-Nusra), 29, 57, 116, 119, 149

Öcalan, Abdullah, 64
Orientalism 5, 47–8

Palestine, 14, 66–7, 122, 124
    Handarat camp, 66–7; Liwa al-Quds (the Jerusalem brigade), 67; al-Nayrab camp, 66–7; Palestinians, 55, 66–7; Yarmouk camp, 66, 158
Palmyra City, 2–3
political parties,
    Arab Socialist Party (led by Akram al-Hourani), 12, 121–2, 124; Baath Party, 10, 11–12, 13–14, 16, 20–1, 49, 66, 97, 100, 104–6, 120, 123, 125, 129, 131; Democratic Arab Socialist Party (led by Jamal al-Atassi), 22; Democratic Union Party (PYD), 31, 59, 64–5, 154; Kurdistan Workers' Party (PKK), 64; Muslim Brotherhood, 16, 21–2, 49–51, 56, 63, 70; Nasserists, 11, 12, 21; National Bloc, 104, 123; National Democratic Rally, 22; People's Party, 122–3; Syrian Communist Party (SCP), 11, 104, 105, 121, 123–4; Syrian Communist Party (Political Bureau led by Riad el-Turk), 22, 39
politics
    of bread, 126, 127, 135–42; of dignity, 108, 109, 110, 113; of life, 4, 6, 7–10, 65, 157; of death, 4, 7–10, 25, 26, 29, 30–1, 37, 38, 67, 155, 157–8, *see also* necropolitics
Postol, Theodore, 2
Power,
    biopower, 25; infrastructural, 135, 138, 142; necropower, 25–7, 32, 37; vertical, 32, 73–78
prison, 5, 10, 23, 28, 33, 37–41, 60; 107, 111–2
    doctor, 34–5; Manbij, 116, 153; Palestine branch, 33; Sednaya, 38; 56; Tadmur, 3, 38, 39
prisoners, 23, 34
    body, 30; Kurdish, 65; massacre, 3, 33; torture, 34–5, 36, 56

al-Qaeda, *see* al-Nusra
Qunaytara, 13

ar-Raqqa, 61, 63, 133, 141, 144, 149
revolt, 1, 4, 6, 8–10, 24–37, 41, 43, 45, 66, 67, 71, 95–9, 105, 120, 134
    Aleppo (2011), 52, 53; anti-colonial, 37, 46; Arab revolts, 1, 4, 6, 24; Druze (2004), 63; Great Syrian Revolt (1919–1920 & 1925–1927), 6, 99–105, 106; Manbij, 107–19, 137–42, 144–56; Mount Lebanon (1860), 45; Muslim Brotherhood (1960s & 1970s), 21, 50; urban, 49–52, 132
Revolutionary Council (RC), *see* Manbij
Revolutionary Court, *see* Manbij
Robinson, Piers, 2

Saleh, Yassin Haj (writer and former political prisoner), 40

Samuel, Ibrahim, (Syrian novelist), 38
sectarianism, 56–9, 95, 102–3, 106, 107, 117–9
Shabiha, 53, 61–2, 63, 70–1
  Berri clan, 30, 63, 70, 72; Kurdish, 71; Lijan Sha'bia (popular committees), 70; Liwa Mouhamad al-Baqir, 71; Liwa al-Quds (the Jerusalem brigade), 67; Martyrs of St. George Brigade, 59; Syrian Christian Resistance, 59; Warriors of Christ's Aleppo, 59
Schmitt, Carl, 17–8
security branches
  hospital 41, 34–5; military intelligence, 66; mukhabarat, 60; Palestine Branch (Far Falistin), 33, 66; Syria Electronic Army, 57
Shehada, Bassel, 115
Sheikh Maqsoud, see Aleppo
Shia, 31, 57, 71
al-Shishakli, Adib, 37
sniper, 29, 31, 32, 34, 35, 41, 43, 69–70, 73–8
  see vertical power
social class, 14, 43, 46, 60, 102, 103, 130
  bourgeoisie, 13, 27, 49, 50, 52, 104, 109, 122–3, 131; capitalist, 77, 124, 131; comprador, 110; elite, 26, 38, 52, 53, 77, 97; feudal, 122, 124; industrialist, 122, 123, 124; landowner, 11, 50, 52, 104, 121–4, 125, 129; merchant, 44, 45, 53, 70, 102, 105, 121, 123, 125–6; middle, 46, 57, 96, 101, 121, 122, 123, 129–30, 146; neo-liberal, 115; peasant, 12, 51–2, 96, 102, 120, 121–32, 134–5, 142; poor, 38, 52, 53, 60; popular, 101, 110; rich, 131; ruling, 99, 122, 123; rural, 50, 52, 57; upper, 64; urban, 46, 49; working, 11, 54, 61, 129
state of emergency, 10, 13–20, 21, 23–5, 36, 41–2
Supreme State Security Court (SSSC), 15

Taftanaz, 109
Taj al-Din, Shaykh, 121
Tammo, Mashaal (Kurdish leader), 31, 65, 115
Tansiqia (pl. Tansiqiat), see committees
tribes, 62–4, 75, 100, 145, 148
  al-Berri, 30, 63, 70, 72; Busha'ban, 63; Mardini, 71
el-Turk, Riyad, 22, see political parties

Umran, Muhammad, 12, 14
Unified Judicial Council (UJC), see legal system
United Arab Republic (UAR), 11–12, 14, 21, 124
Urbicide, 8, 43, 54–6, 67, 71, 72, 73, 76
  medical facilities, 2, 32–3, 34, 35, 62, 73; military urbanism, 55, 78–9; spatial tactics, 55
Useful Syria, 31–2, 68, 95, 136

war, 78–9, 136
  Arab-Israeli 1967, 13, 50; World War I, 100–1; World War II, 121; Yugoslavia civil war, 74

al-Yarmouk (Palestinian camp), 29, 66, 158
Young Turks, 100

al-Zaim, Husni, 37

www.ingramcontent.com/pod-product-compliance
Lightning Source LLC
Chambersburg PA
CBHW032036290426
44110CB00012B/828